YAM

in WEST AFRICA

YAM

in WEST AFRICA

Food, Money, and More

Felix I. Nweke

Michigan State University Press | *East Lansing*

♾ The paper used in this publication meets the minimum requirements
of ANSI/NISO Z39.48-1992 (R 1997) (Permanence of Paper).

Michigan State University Press
East Lansing, Michigan 48823-5245

Printed and bound in the United States of America.

22 21 20 19 18 17 16   1 2 3 4 5 6 7 8 9 10

Library of Congress Control Number: 2015939353
ISBN: 978-1-61186-187-7 (pbk.)
ISBN: 978-1-60917-474-3 (ebook: PDF)

Book design by Charlie Sharp, Sharp Designs
Cover design by Shaun Allshouse, www.shaunallshouse.com
Cover image of yam mound making, Abakaliki, Eastern Nigeria is used courtesy of Louise Fresco.
Cover image of ware yam market scene, Kintamkpo, Ghana is used courtesy of Felix I. Nweke.

Michigan State University Press is a member of the Green Press Initiative and is committed to developing
and encouraging ecologically responsible publishing practices. For more information about the Green Press
Initiative and the use of recycled paper in book publishing, please visit *www.greenpressinitiative.org*.

Visit Michigan State University Press at *www.msupress.org*

To

CARL K. EICHER
With gratitude for over fifty years of tutelage.

To

My grandchildren:
OBIDIKE OBANU OFOEDU NWEKE-ONWEZI,
DALUCHI SOMGOLIE ONYEYILI NWEKE,
IZUNDU CHIKELU OMEOCHA NWEKE
(April 28, 2011 to January 15, 2013),
and
CHIDERA AKUNNA NDIATU NWEKE
*For the lesson that life and death are two sides of existence;
to have one is to have the other.*

"You come from a family of yam farmers; they are the salt of the earth to you. They represent the heart of Nigeria. You joined the army [of researchers] to protect them."

—NNEDI OKORAFOR, *Lagoon*

Contents

Foreword

Werner Kiene

N o matter whether one is a development practitioner, a researcher, or a student, one cannot help but be amazed by yam—the most African of African crops. This book admirably succeeds in weaving a broad canvas of information about yam's significance in West Africa and by the same token for Africa's food security as a whole. The book deals with the unique aspects of yam as it is grown, marketed, and consumed. On top of this, the book demonstrates in a fascinating way how culture is intimately related to what people eat and what foods they produce.

Yam in West Africa is a timely and important book for Africa's "yam belt," a region stretching from the coastal countries of Côte d'Ivoire, Ghana, Togo, Benin Republic, and Nigeria to Cameroon and parts of the sub-Sahel countries including Mali, Burkina Faso, Niger, and Chad. The book's central thesis is that West African consumers would love to eat more yam if it would reach the market at more affordable prices and producers would grow more yam if the crop could be produced profitably at lower costs.

Yam is indeed the soul of West African cuisine. The book provides a vivid account of the importance that culture and tradition play in the production and consumption of food. Too often this is overlooked in deciding on food policies and related research and action.

In spite of yam's economic, nutritional, and cultural values, the author documents a history of neglect and almost perverse policy bias against this food crop that has been so important in the lives and traditions of so many West Africans. The recommendations flowing from the book point into promising R&D directions that will make yam more affordable to consumers and more profitable to farmers. Of particular importance for policy makers, researchers, and students are the sections that document the challenges and opportunities of yam seed production, yam disease control, and yam market improvements. But sections on labor, crop improvement research, technology transfer, policy, gender, postharvest matters, fuels of consumption, and cultural values are equally illuminating.

Only recently have larger R&D efforts been devoted to yam, and it should be noted that Professor Nweke—as he did some years back with cassava—has played a key role in increasing our understanding of yam-based farming systems and supporting policies. As a former fellow graduate student with Felix Nweke at Michigan State University in the early 1970s, I recall that it was he who introduced me and some other fellow graduate students to this amazing crop. He has been on this mission ever since then, and countless researchers and policy makers owe him thanks for his tireless efforts in raising awareness of yam's neglected potential for West African food security.

Yam in West Africa is a social science book on yam—the first of its kind. However, the book is also a testimony to Professor Nweke's profound personal connection and emotional attachment to yam. In the book's preface, he gives us a touching glimpse at that personal relationship and shares with us memories of those who have supported him along the way: from his childhood in a West African village to becoming an internationally recognized scientist, yam has always been part of his journey. It is about time that this book was written.

Preface

My interest in writing this book was sparked by study of yam consumption patterns in West Africa. In December of 2012 and January of 2013 I traveled across West Africa by bush buses, talking with and observing yam producers, distributors, and consumers. I found the problems of low technology, high production costs, and high consumer prices in the yam food sector and the deep poverty among the yam-producing households striking. Informal discussion of these issues with friends and family members, including Chief Saadallah Saklawy, my kindred spirit, Professor David Ifufudu, my son's father in-law, and Professor Ajuruchukwu Obi, my former student, confirmed that the yam story needed to be told.

The yam story as it is told in this book is based on research conducted in recent years. But the book also conveys insights and experiences that I assimilated throughout my whole life from the mid-twentieth century when I grew up at Ukpo, Dunukofia in eastern Nigeria. From early childhood, my life and that of my extended family revolved around producing, eating, and often marketing yam and cassava. Those experiences formed me and are at the base of my professional career and my interest in the subject of this book.

Acknowledgement and thanks are owed to a large number of people, but I would like to pay special tribute to four generations of my family, from whom I learned a lot and for whom yam was, and still is, an important part of their lives. I met only my maternal grandparents, Onyeyili and Mgboye Oji, who contributed considerable influence to the formation of my worldview because they raised me close to their hearts. Members of the generation after that of my grandparents include my parents, Nweke and Joy-Nweke Akpuocha, and my uncles and aunts represented by Andrew Onyeyili, Okafor Onyeyili, Dickson Onyeyili, Nwafor Akpuocha, Mgbafor Ofogeli, Eugenia Onyeyili, and Ekenma Nwafor-Akpuocha. My generation is made up of my wife, Nwogo Achebe-Nweke; my siblings, Edwin Chikelue Nweke, Catharine Nwakaego Nweke-Anyanwutaku, Kenneth Nweke, Dora Afue Nweke, and Agnes Ngozi Nweke-Muokwua. That generation also includes my cousins represented by Fred Onyeyili, Joseph Nwamulunamma, Christopher Nwafor, John Umeadi Anadebe, and Simeon Nwafor. The fourth generation consists of my children, Nnake Nweke, Arize Nweke, and Nnedi Ifudu-Nweke; and several nephews and nieces represented by Bertha Anyanwutaku-Okonkwo, Chudi Chikelue, and Kingsley Osondu Ezeaku. Yam continues to be central in the food and culture of my children's generation just as it was in my grandparent's generation. This connectedness now reaches my little grandchildren who live in Maryland, United States, and frequently ask for "yam French fries" for dinner.

Of course, I owe thanks to members of my generation beyond my family. In Ukpo, Dunukofia on reaching the age of eighteen to twenty-one years, people organize themselves in cohorts of persons born within a three-year interval. Each cohort meets regularly, engages in community development projects, and provides support for members in distress. As a mandate at a funeral of a colleague, members contribute a tuber of yam from among the largest available in the market, as a symbolic gesture for sustenance of the family during the period of grief. Igbo people say proverbially that children do not starve in the year of their father's death because of the support group contributions. Of course, from the beginning of time, to the Igbo man, food was yam. My age cohort is Uzodinma Age Grade—I hosted the meeting held on December 31, 2013, with more than fifty members in attendance.

I crossed paths with several influential teachers and schoolmates and I acknowledge with thanks the impact they had on my thinking. At primary school, my most influential teachers were John Achebe who made us grow yam in the school garden, Augustine Esedo, Anthony Onyeka, John Mezue, and Emmanual Onuegbu. My closest primary schoolmates were Samuel Ugoezue, Paul Okeke, Mabel Uwaezuoke,

Maxwell Nwobo, Eunice Okeke-Okpokwasili, Ernest Okeke, Christopher Ofogeli, Nnaemeka Ikpeze, Simon Nzeadi, Christopher Ofogeli, Eric Omejilichi, and Isaac Oliobi. At the Community Secondary School, Nnobi, my favorite teachers were Elisha Etudo, Ephraim Dalah, Chinwuba Ezeaku, Okwuchukwu Ezeaku, Shedrack Ifedili, Raphael Igwedibia, Okechi Amu-Nnadi, Lambert Onuora, and C. J. Ogbuka. Eugene Iloansi was the secondary schoolmate who inspired me most.

In undergraduate school at the University of Nigeria, Nsukka, my encounter with Professors Bede Okigbo, Fred Ezedinma, Warren Vincent, Glenn Johnson, Dale Hathaway, Martin Billings, and Carl Eicher consolidated my interest in pursuing a career in agricultural development. It was at the University of Nigeria in 1963 that the lifelong working relationship with Professor Carl Eicher, the foremost contemporary expert on African agricultural development, began, and it was with joy that in 2013 we celebrated the golden jubilee of that relationship. This book could not have been written without the foundation that was laid by my association with Professor Eicher, and I am grateful for the handwritten note that Carl addressed to me on March 11, 2014, wishing me to stay in good health and urging me to continue the journey we started fifty years earlier. Carl knew that this book was in preparation but did not live to see it published because he passed away on July 5, 2014, while I was away in Tanzania pursuing our mutual journey as he requested.

Acknowledgement is also due to my students and my professional colleagues with whom I collaborated on issues that relate to this book. As professor of agricultural economics at the University of Nigeria, Nsukka, I led research projects such as the following:

1. "Bases for Resource Allocation in Yam Cropping System in Eastern Nigeria" in collaboration with Fred Winch of the International Institute of Tropical Agriculture (IITA), Ibadan
2. "Consequences of Crop Stereotypes along Sex Lines in Yam Cropping System in Eastern Nigeria" in collaboration with Werner Kiene of the Ford Foundation
3. "Regional Market Demand for Yam in Eastern Nigeria" in collaboration with David King of the International Development Research Centre, Ottawa

Talented students who contributed important insights in these studies include Jude Njoku, Charles Asadu, Boni Ugwu, Ajuruchukwu Obi, Donatus Ibe, Godwin

Asumugha, Lambert Eluagu, and Eugene Okorji. Fellow professors and my mentors at the University of Nigeria including Patrick Ngody, Dennis Ekpete, Walter Enwezor, Maxi Ikeme, Frank Ndili, James Nwoye Adichie, and Chimere Ikeoku were supportive of my work.

It is with great appreciation that I reflect on my collaboration with IITA. In 1987 I was hired at the IITA as yam-based farming systems economist in the institute's Resource and Crops Management Program. One year later in 1988 I was reassigned as the project leader of the Collaborative Study of Cassava in Africa (COSCA) studies; the COSCA studies included collection and analyses of information on yam, which is produced in the same agroecologies as cassava in West Africa. In these studies I collaborated closely with John Lynam of the Rockefeller Foundation, John Strauss of Michigan State University, Andrew Westby of University of Greenwich, Louise Fresco of Wageningen University, Hans Rosling of the Karolinska Institute, and Eric Tollens and Muamba Tshiunza of Katholieke Universiteit Leuven. At the IITA, Dunstan Spencer, Margret Quin, Larry Stifel, San Ki Han, Lukas Brader, John Takway, Henk Mutsaers, Louise Fresco, Peter Ay, and others were supportive of my work. They all deserve thanks.

Progressive leadership provided in my hometown by Walter Eze, Robert Eze, Paul Okeke, Herbert Okonkwo, Chinyere Okunna, and others at the town level, and by Edwin Nweke, Patrick Okoye, Francis Okafor-Etoh, Raphael Akude, Louis Nwankwo, Paul Amaonye, and others at various levels of my extended family allowed me peace of mind as I pursued my career away from home. Michael Eze, Arthur Eze, Paul Okeke, Bertram Okpokwasili II, Mary Eze-Ekwue, Samuel Ifeacho, Emma Nwoyenma Nwankwo, Emmanuel Ezekwesili, Chukwuemeka Nwoye, Sunday Eseaku, Sunday Ukonmadu, Ifeanyi Omeokachie, Anayo Ejem, Ofodile Okafor, Eric Okoye, Gideon Nweke Otubelu, and Godfrey Paul Mary Okoye are among many of the people from my hometown who provided financial, material, and other forms of support at various times when I needed help.

This book draws on information generated in two farm-level surveys, the Yam Consumption Patterns in West Africa study funded by the Bill and Melinda Gates Foundation and the baseline survey of the Yam Improvement for Income and Food Security in West Africa (YIIFSWA) project at IITA. The work also benefited from various YIIFSWA Annual Progress Review and Planning meetings in which I participated. Lists of team members of the surveys and of participants in the YIIFSWA meetings are provided in appendix 1. I extend my appreciation to them and to those who supported us in those endeavors.

In several instances, Issahaq Suleman helped to secure, verify, or clarify information on Ghana. Werner Kiene and Chris Okonkwo read the draft manuscript. Their comments resulted in considerable improvement. From time to time, as I worked on the book drafts in Nigeria, I went on retreats to escape frequent interruptions by visits of numerous friends and family members. On such occasions, depending on where I went, Paul Okeke provided free board and lodging in his Lone Palm Hotel, Asaba; Saadallah Saklawy paid for my stays at the Mount Pleasant Hotel in Ibadan.

I am grateful for my association with the African Studies Center at Michigan State University. Special thanks go to two former directors of the center, David Wiley and James Pritchett, for providing me with a position of visiting scholar.

This book is a result of efforts of many people over a period of close to seventy years. Obviously, many more people deserve to be acknowledged but are missed; to them I apologize. As the author I take responsibility for all shortcomings due to decisions I made.

Plan of this Book

This book is in fourteen chapters. Chapter 1 is an introduction to yam and is particularly important for readers outside West Africa who are not familiar with the crop. Chapter 2 is a characterization of the contexts in which yam is produced in West Africa with emphasis on the poverty levels in yam-producing communities. Chapters 3, 4, and 5 focus on the analysis of the three major challenges in the yam crop sector: chapter 3, the high cost of production labor; chapter 4, the high cost of seed yam; and chapter 5, yam plant pests and diseases.

Chapter 6 focuses on yam breeding, and chapter 7 analyzes the prospects and impediments to the diffusion of hybrid yams officially released to farmers in Nigeria beginning in 2001. Chapter 8 illustrates the periphery situation of the yam in national food policy programs in West Africa and attributes this unfavorable situation to low technologies in the yam crop sector. Chapter 9 deals with the important issue of gender and concludes that the common reference to yam as man's crop is a stereotype.

Chapter 10 takes up the whole gamut of postharvest issues, including how yam storage technology varies with place but not with time and how the marketing channel is short of rural assembly because of yam pests and diseases. Chapter 11 focuses on analysis of the trade in West African yam.

Chapter 12 establishes that consumer income, relative prices, and yam food preparation technology are significant variables in fueling yam consumption in West Africa. Chapter 13 illustrates the role of yam in cultural rites practiced in several yam-producing areas, such as the marriage rite in eastern Nigeria, and raises the question of the implication of ceremonial use for yam technology development. Chapter 14 is a synthesis chapter that ties together the various analyses of progress and challenges in yam technology development and transfer in West Africa.

Introduction

Yam has been sidelined in national food policy programs in West Africa and ignored by African development agencies for much too long a time because it is considered to be a complex crop that would not respond positively to investment in research and development (R&D). Yam, a commodity that is appreciated by so many people, a commodity that is part of the traditional diet of many and is central in cultural rites that are important for the existence of the people in producing areas individually and communally, has for so long been without appropriate R&D attention.

For instance, the African Development Bank, Africa's premier development institution, provides loans and grants for R&D on food staples in fifty-four member countries, but yam is not included. Similarly, Nigeria's minister of agriculture, Dr. Akinwumi Adesina, introduced a transformation agenda for the purpose of promoting nine commodities; all major staples are included in this list, but yam is not (Nigeria 2011). The International Fund for Agricultural Development (IFAD) is an exception. In the 1990s IFAD funded a program for the expansion of root and tuber crops in West Africa. The yam component of the program was based on the minisett seed yam technology, which, as will be seen in later chapters, had a limited degree of adoption (RTIMP 2009).

What are the obstacles that need to be cleared in order to rescue and clean up the precious baby, that is, yam? What measures need to be implemented so that the commodity will assume its proper place as Africa's second world crop after cassava?

The continent's contributions to global supplies of grains are modest: maize, about 5 percent; rice, 3 percent; and wheat, 3 percent in the late 2000s (FAOSTAT).[1] However, Africa is the lead player in terms of the production of yam and cassava: more than 90 percent of global yam production and 50 percent of cassava production happens in Africa. Unfortunately, as will be demonstrated in the following chapters, yam is produced at high costs because of the low level of technologies used in both production and postharvest handling. This book contends that yam could and should play a much more important role in Africa's existing and future development efforts if necessary measures are put in place to develop the yam food sector.

West Africa is a region with an estimated population of 350 million people in the year 2014 with annual compound growth rates of around 2 percent (FAOSTAT). Most of these populations are dependent on arable crop agriculture and derive large percentages of their food calorie intake and cash income from root and tuber crops, mostly cassava and yam. This suggests that millions of Africans are negatively impacted by the low productivity of yam cultivation, which results in low household incomes. By the same token, high prices have negatively impacted incomes of consumers and the diversity of their diets.

Shifting the Focus

An important objective of this book is to draw attention to the needs and opportunities there are in the technological development of yam and to the benefits such development would produce in terms of improved producer and consumer incomes, food availability, and nutrition. Inspiration should be taken from the success that was experienced from increased research financing in the case of cassava production. After decades of colonial and postcolonial focus on cereals and grains such as wheat, rice, and maize, which were previously considered "delicacies" or "glamor crops" in West Africa, cassava began to receive attention some forty years ago (Johnston 1958, 226; Jones 1972, 28).

The success achieved in the technological development of cassava eventually led to expanded R&D funding following the diffusion of high-yielding, mosaic resistant Tropical Manioc Selection (TMS) varieties developed by the International Institute of Tropical Agriculture (IITA) and the Collaborative Study of Cassava in Africa (COSCA) studies that unveiled the crop's potential as a powerful poverty fighter on the continent (Nweke, Spencer, and Lynam 2002). This book is based on the belief that a similar success story could be produced if appropriate R&D attention is devoted to yam by West African political leaders, policy makers, researchers, and international donor organizations.

Why Write a Book on Yam in West Africa?

The coastal countries of West Africa from Cameroon to Côte d'Ivoire house the yam belt of the world, which is a subset of cassava agroecologies of West Africa. Carbohydrates are the main food value of both yam and cassava; the dilemma is that consumers pay premium price for yam when they can get the same food value from cassava at a lower cost. This book is an attempt to contribute an answer to a question often asked in R&D circles—namely, what does yam do for consumers that cassava does not? The book goes further to identify the reason why the yam is sidelined in national food policy programs and the causes of the high cost of yam. The book will also suggest necessary measures to be taken in order to make the yam price competitive with cassava.

In November 1961 an unpublished nineteen-page report, "Review of Yam Research in Nigeria 1920–1961," by A. W. Waitt, a British colonial botanist, was issued by the Department of Agricultural Research of the newly independent Federal Government of Nigeria (Waitt 1961). Waitt's report is important because it is the earliest documented information on yam research in Nigeria and is widely referenced by yam scientists in the country today. The report is a documentation of scientific information available on yam in Nigeria up to 1961.

In 1967, a landmark book on yam was published by D. G. Coursey, another British colonial botanist who worked in both Ghana and Nigeria up to the dawn of Africa's independence (Coursey 1967). Coursey's book, *Yams*, which is useful to yam researchers mostly in terms of its ethnobotany content, includes a chapter on the agriculture of yams and another on yam festivals in eastern Nigeria. In 1982, J. Miege and S. N. Lyonga edited the book *Yams*, a collection of papers covering plant improvement, agronomy and taxonomy, plant protection, postharvest biology and technology, and strategy for reducing labor requirements (Miege and Lyonga 1982).

In 1993, L. Degras also published a book on the crop, *The Yam: A Tropical Root Crop*, which is good reference material on the water yam in Côte d'Ivoire (Degras 1993). In 1998, G. C. Orkwor and others published *Food Yams: Advances in Research*, a collection of high-quality scientific papers on yam in West Africa that should be of interest mostly to biological scientists (Orkwor, Asiedu, and Ekanayake 1998).

Bringing the Complexities of Yam to a Wider Audience

This book is aimed at a broader audience; the information conveyed in it is designed to be of value to a wide cross section of yam R&D practitioners including yam researchers, West African agricultural policy makers and political leaders, international donor organizations, and certain yam entrepreneurs, such as yam export and import traders.

The book's focus is on three of six cultivated yam varieties—namely, white, water, and yellow yams that are most commonly produced, marketed, and consumed in West Africa as well as exported to several countries worldwide. The book lays emphasis on three bottleneck effects in the development of the yam crop sector—namely the high cost of labor, the high cost of seed yam, and the yam pest and disease problems—highlighting achievements, outstanding challenges, and research efforts to break these bottlenecks. The importance of the widely produced ceremonial yam in West Africa, perhaps as an impediment to yam technological development, is of significant interest in the present book. For this purpose the book highlights, perhaps for the first time, a less measurable value of yam beyond food and money.

Yam in the National Economies of Nigeria and Ghana

Although yam is produced in many parts of the world, West Africa is the subregion of the main concentration of its production; the white yam is indigenous to West Africa (Coursey 1967). From 2006 to 2010, approximately fifty million tons of yams were produced annually worldwide (FAOSTAT). West Africa contributed more than 90 percent and the rest of the world less than 10 percent of the global yam production. Within West Africa, yam production is concentrated in Nigeria and Ghana. These two countries produced 75 percent of the global yam supply from

2006 to 2010, Nigeria, 65 percent and Ghana, 10 percent, leaving the rest of West Africa providing the balance of 15 percent of the global yam supply.

In West Africa, yam is one of the principal food crops produced in the subhumid agroecologies, which corresponds to the middle zones of the coastal countries of the subregion from Cameroon to Côte d'Ivoire, the so-called yam belt of the world. North of that zone, yam production pales to a marginal scale in nonhumid agroecologies; south of the yam belt, production also pales to a marginal scale in the humid forest zone.

Yam is a major source of cash income for millions of producing households because it has a high market demand and it is easily exchanged for cash in rural and urban markets. In cassava-producing areas of Nigeria where yam is only a secondary crop, yam contributed 18 percent of household food crop cash income, second to cassava, which contributed 22 percent (Nweke, Spencer, and Lynam 2002).

FAOSTAT records yam exports from Ghana where the crop is an increasing source of foreign exchange. Yam is exported from Ghana to several other countries in West Africa where production is insufficient to satisfy the consumption needs of the population. Yam produced in West Africa is widely available in Europe, in North and South America, and in parts of Asia where it is consumed by increasing immigrant African populations.

Sources of Information for the Present Book

The Yam Consumption Patterns in West Africa study (hereafter yam consumption study), commissioned by the Bill and Melinda Gates Foundation (BMGF) and executed from November 2012 to February 2013, is the original work done for this book. The yam consumption study was conducted in four purposely selected countries—Burkina Faso, Ghana, Mali, and Nigeria—representing West Africa. In 2010, these four countries accounted for more than 70 percent of the population of West Africa and more than 85 percent of the West African yam supply (FAOSTAT). Nigeria and Ghana are respectively the largest and second largest yam producers worldwide; Mali and Burkina Faso are marginal yam-producing countries.

In each representative country, a nationwide food consumption survey was conducted. Time and money available for the study were tight; there were barely four months from November 8, 2012 to February 28, 2013 to complete the survey, and the financial budget for the nationwide food consumption survey was zero. Without a nationwide food consumption survey, analyses of yam consumption patterns would only be speculative and nonfactual because of the scanty information in yam literature for West Africa.

The food consumption surveys were conducted by telephone in all four countries. In each country, a number of telephone enumerators were engaged. Enumerators interviewed their telephone contacts using a single row structured questionnaire. Information collected included location, gender, income group of the respondent, and number of times the individual respondent ate each of the major staples in the preceding seven days; some of the staples were broken down into specific food products. Location, gender, and income were known to the enumerator a priori because the respondent was a telephone contact. Based on personal knowledge, the enumerator assigned a respondent to a lower, medium, or upper income group. Therefore the income grouping of the respondents is not unique among enumerators. The telephone interview was efficient; a short and simple telephone interview guaranteed wide geographical coverage in a cost-effective manner.

In the major yam-producing countries, that is, Nigeria and Ghana, farmers were interviewed in groups in different yam agroecologies. A group consisted of a minimum of ten farmers widely ranging in age, including the oldest in the community. In Nigeria, the farmer group interviews were conducted in Otuocha, which is known for seed yam and ceremonial yam production along the River Niger basin, in Zaki Biam in the derived savanna agroecology, and in Shaki, north of Ibadan, which is celebrated as the largest source of traditional yam flour called *amala*. Farmer groups were also interviewed in Kintamkpo in central Ghana and in the Tamale area in northern Ghana. Information was sought from the farmers on yam production, harvesting, and postharvesting handling technologies, and on the uses of yam apart from selling and home consumption.

In a major yam market in each study country, a survey was conducted to perfunctorily assess the volume and determine, through the interview of merchants, the origin of yams available for sale. In Ibadan, Nigeria, processed yam products, namely *amala* (traditionally prepared yam flour) and *poundo* yam (industrially prepared yam flour) and their substitutes, namely *semo* (industrially prepared flours of grain) and *gari* (granulated cassava product) were purchased to determine their prices at retail value. The retailers were interviewed for their assessment of purchaser preferences for those products. Prepared food made of yam, namely *foutou* (pounded yam) and its substitutes, *to* (maize meal) and *riz* (rice meal), were bought in a popular restaurant in central Ouagadougou, the capital of Burkina Faso, to determine prices. In the market in Kumasi, Ghana, yam, fresh cassava roots, and maize grains were purchased to determine their prices at retail value.

All travels for interviews and observations within and among the four countries studied were by road. Road travels by bush bus were undertaken from Lagos through Lomé, Ouagadougou, Bamako, Tamale, Kumasi, and Accra and back to Lagos. This intimidating effort was necessary to observe the movement of yam by merchants and individual consumers through informal channels because of the deficiency in recorded data on yam trade in West Africa. The effort was also necessary because the importance of market access infrastructure for yam trade demands an assessment of the quality of intra- and interstate roads and other transport facilities, even if cursorily.

Some original farm-level data collected by IITA scientists in the context of the Yam Improvement for Income and Food Security in West Africa (YIIFSWA) project is used in this book. The bulk of the information presented in this book came from extensive literature review; in this respect, the proceedings of the triennial symposium of the International Society for Tropical Root Crops–Africa Branch and the annual reports of the IITA are particularly rich. In fact, those two sources are treasure mines of information on yam in West Africa. Additional resources are farm and home-level studies on production and consumption of the yam and other staples in West Africa that are reported in literature. FAOSTAT is the source of information on the yam trade, population, production, consumption, and prices for yam and other staples over time.

CHAPTER 1

Yam Primer

Y am is a complex crop in several ways. First, it is frequently confused with other crops that go by the same name outside of West Africa. Second, its production method does not lend itself to comprehensive analysis because it varies significantly in different agroecologies within the West African yam belt and in other yam-producing areas of the world. In some parts of West Africa, production methods for yams of different sizes vary significantly enough to consider each size category of yam a different crop. Similarly, variations in the uses made of yams of different sizes lend support to considering the yams in different size categories as different crops. Third, in the West African yam belt, the yam's production cost is extremely high, which foreshadows the potential food and income-generation values of the crop. Fourth, in most of the yam-producing communities in West Africa, yam plays important cultural roles that set it apart from other food crops in the communities.

What Is Yam?

D. G. Coursey defined the yam by elimination of what is not yam but is called yam in different parts of the world outside West Africa (Coursey 1967). In the United States of America, yam is commonly understood to be sweet potato. Yams are also often confused with edible aroids such as cocoyams, taros, etc. outside of West Africa. In India, the elephant yam is an aroid, which is related to the cocoyam but not to yam. The word "yam" has also been used for the arrowroot and for several other edible starchy roots, tubers, or rhizomes grown in the tropics. Some leguminous plants that have swollen edible roots are described as yam beans.

In the sense that the word is used in this book, all yams are members of the monocotyledonous Dioscoreaceae family; virtually all belong to the genus *Dioscorea*. The genus *Dioscorea* is made up of some six hundred species, but only about ten of them produce edible tubers (Ene and Okoli 1985).

The majority of *Dioscorea* that are of economic importance produce one or more tubers underground, which are often renewed annually; few of the *Dioscorea* such as aerial yams, which produce tubers above the ground, are exceptions. Above the ground, the growth, which is usually annual, consists of twining vine-like morphology that requires support from neighboring vegetation or a stake (Waitt 1961). Flowering is rare, and male and female flowers that are tiny are most often on separate plants. When flowering occurs, setting of fertile seeds is even rarer than the flowering. The seeds are small, light, winged, and dispersible by wind (Coursey 1967; Ene and Okoli 1985).

Yam Size Categories in Eastern Nigeria

For practical agronomic purposes, researchers categorize yam within a variety into two: seed yam and ware yam.[1] These differ in tuber size, production methods with respect to the size of the seed yam, mound in which the yam is planted, stake provided for the yam to twine on, and also with respect to planting date, soil fertility, and the yam stand density. In eastern Nigeria, farmers have three categories of yam: seed yam, table yam, and ceremonial yam. In the Otuocha area of eastern Nigeria, farmers have a name for each category: *awa ji* for seed yam and *nnukwu ji* for ceremonial yam; the residual, that is, not seed yam and not ceremonial yam, is table yam.

Within a variety, the three yam categories differ in tuber size; ceremonial yam is the largest and seed yam is the smallest of the three. Production methods differ considerably with respect to the size of seed yam sown, the size of the mound in which the yam is planted, and the size of the stake provided for twining. Ceremonial yam is produced by farmers with the largest seed yam, mound, and stake, while seed yam is produced with the smallest mound and materials. Production methods also differ with respect to planting date, soil fertility, and stand density. Ceremonial yam is planted by farmers at the beginning of the planting season and in the most fertile soil, but at lowest stand density. Seed yam is planted latest in the planting season in the lowest fertility soil, but at highest stand density. Practices in the production of table yam are intermediate between practices in respect to ceremonial and seed yams. Recognition of size categories by yam researchers will help in targeting R&D interventions in the yam crop sector.

Growing yam size categories as independent crops is common only in eastern Nigeria and surrounding areas. Elsewhere in the West African yam belt yam is sorted into size category at harvest of a field in which yam is grown as one crop. In a yam crop field, tubers produced vary in size; depending on microenvironmental variations within the field, such as variations in soil fertility and in soilborne yam pests and diseases, different plants produce tubers of different sizes. In addition, occasionally a yam plant produces multiple tubers of varying sizes. At harvest, tubers are sorted by size: small ones for seed yam, medium ones for table yam, and the very large ones for ceremonial yam.

Categorization of crop by size is unique to yam among food crops; it is one of the numerous sources of complexity in understanding yam production and utilization. Although yam of any size category is edible, different size categories of yam have different uses besides consumption. Before ceremonial-sized yam ends up on the dining table, it must have performed an intermediate function of ceremonial use, while seed yam is more valuable as seed than as food. Therefore, different size categories of yam are also different crops in terms of utilization.

Yam Cultivation Methods in Eastern Nigeria

Nigeria's yam belt spans over six major agroecologies, which means there are at least six different yam production systems. The yam production system in one of the six agroecologies in Nigeria's yam zone—the forest regrowth agroecological zone—is discussed in this section because there is more availability of information. North of the forest ecology is the derived savanna ecology, and north of that is the Guinea savanna (hereafter the savanna) ecology, all within eastern Nigeria.

Eastern Nigeria, the area south of the River Benue and east of the River Niger, stretches from the humid forest to the subhumid Guinea savanna ecological zones; most of this area falls within the yam belt of Nigeria (Nweke et al. 1991). A large section of eastern Nigeria is part of the catchment area of the River Niger, which is considered to be the center of origin of the white and yellow yams (Coursey 1967).

In this section, emphasis is on yam cultivation methods in eastern Nigeria illustrated with the methods as practiced in the Ukpo, Dunukofia community in the Otuocha area. The reason for this is that in West Africa there is a wide diversity of yam cultivation systems with respect to different practices of intercropping, staking, and land preparation; the diversity is determined by differences in weather and soil conditions (Akoroda 1992; Waitt 1961). Some common practices prevail across all the ecological zones. For example, in all ecological zones, yam is propagated vegetatively with the tuber, that is, the crop. This practice is a source of high yam production costs and prices. In virtually all yam ecologies in West Africa, the crop is planted in a mound seedbed; one of a few exceptions is the narrow bed of the River Niger where yam is planted on the flat land because the soil, which is made up of mostly alluvial deposits, is deep and well drained. Some other practices are specific to ecological zone. For example, staking varies in elaborateness from very elaborate in humid forest areas to no staking in some savanna ecologies.

In most years in eastern Nigeria, as in the rest of West Africa, low insolation caused by dust-borne cool and dry wind from the Sahara desert overcasts the atmosphere and blocks most of the direct rays from the sun. This prevails from November to February and is described as the harmattan season. Insolation is lower in the forest than in the savanna ecology during the harmattan season and also during the rainy season, which is usually from May to October each year, creating a greater need for staking yams in the forest than in the savanna ecology. Water balance is higher in the forest than in the savanna agroecology, although difference in distribution is marginal; 90 percent of the total annual rainfall usually occurs in six and a half months in the forest and in six months in the savanna ecology. In the forest ecology yam is planted beginning with ceremonial yams in February in hydromorphic soils and ends about June each year with the planting of seed yams. Most of the forest soils are deep but not well drained.

The yam consumption study researchers investigated the yam production practices in the Ukpo, Dunukofia community in the Otuocha area in eastern Nigeria within the forest ecological zone. The Otuocha area is reputed for seed and ceremonial yam productions, but in Ukpo, Dunukofia, all yam size categories described above are grown. They are planted in mounds and staked; propagation is vegetative with the tuber, that is, the crop.

Crop rotation and land fallow are features of the yam cropping systems in the Ukpo, Dunukofia community, although population and market pressures are relentlessly manipulating these practices. The entire Otuocha area including the Ukpo, Dunukofia community has come under rapid commercialization and population growth because of the area's proximity to Onitsha, a historically major commercial center in eastern Nigeria. Most of the time yam is the first crop after fallow in a rotation cycle.

In the Ukpo, Dunukofia community, yam is almost always intercropped; it is the major crop in an intercrop system that consistently includes cassava, maize, cocoyam, and a wide range of vegetables. These intercrops, which are grown as part of the yam production enterprise, help distribute the high costs of land clearing and seedbed preparation for yam and augment the profitability of the yam enterprise. Additionally, the practice of growing a number of different crops in a field at the same time is convenient in terms of land resource management under rapid population growth and commercialization, which are increasing the value of farmland in the area.[2]

Yam Food Preparation

In West Africa, yam is preferred as fresh tuber for consumption. Fresh tuber can be prepared in several ways such as boiled, fried, roasted, or pounded. Yam food preparation from preprocessed form, namely dried tuber flour, is less common.

In certain respects fresh yam tuber is more convenient to prepare in comparison with other food staples, especially beans, peas, maize, and rice. Boiled yam cooks faster and therefore uses less energy and time than any of the other staples. The downside is that yam purchases for home consumption must be made more frequently than beans, peas, maize, and rice because yam tuber is more susceptible to damage by pests and diseases than those other staples.

Nigeria is the center of a wide variety of yam food preparations; all the forms enumerated above are common in Nigeria. Consumption of dried tuber flour, which

TABLE 1. West Africa: Average carbohydrate, protein, and fat per kg per commodity

COMMODITY	CARBOHYDRATE (KCAL.)	PROTEIN (GM)	FAT (GM)
Yam	985	15.9	1.87
Cassava	909	5.9	1.56
Sweet potato	929	11.6	3.87
Maize	3,103	81.7	86.30
Rice	3,636	734	7.55
Pulses	3,360	220	16

Source: Data from FAOSTAT.

is rare elsewhere in West Africa, is common in certain parts of Nigeria. In Ghana, preparations such as boiled and fried yam are common, but preparation of pounded yam competes unfavorably with pounded cassava, plantains, and pounded cocoyam especially in the southern regions where these other staples are widely produced at lower costs and therefore lower in consumer prices than yam.

In the Otuocha area, roasted yam is commonly prepared and eaten in the field while harvesting yam, especially seed yam. The suitability of this yam meal under field conditions includes the abundance of firewood in the form of used yam stakes. The suitability also includes a simple recipe: roasted yam is often eaten with palm oil and salt or without any condiments. It is appreciated by many in the area because of its burned flavor. Outside field conditions roasted yam is a delicacy in West Africa often prepared and eaten in relaxed situations. Recently in Nigeria, roasted yam has been commercialized as a convenient food prepared along highways and sold to travelers. It can also be found around university campuses for student patronage.

Nutritional Value of Yam

Questions are often raised in agriculture R&D circles in West Africa concerning the nutritional value of yam, especially when measured against cassava, which is reputed as a low-cost source of food calories. These questions raise doubt among African policy makers and international donor organizations regarding the rationale for investment in yam R&D. Their reasoning is that a unit of investment is more profitable, in terms of food calories produced, in cassava than in yam.

Analysis of FAO Food Balance Sheets shows that the main nutritional value of yam, cassava, and sweet potato are food calories. This is because those starchy staples are considerably lower in protein and fats than are pulses and grains (table 1). The margin of difference in calorie content between cassava and yam, which is in favor of yam, is low enough that production cost per unit of calorie is likely to be lower from cassava than from yam since yam is more expensive to produce than cassava. Protein content is considerably higher in yam than in cassava, yet yam is not a major source of protein when weighed against pulses and grains. Similar comparisons are obtained with respect to fats.

In conclusion, cassava is likely to be a cheaper source of calorie than yam. But the fact that yam has a considerably larger amount of protein and perhaps more calories as well as fats per unit weight than cassava is clearly sufficient justification for addressing the underlying causes of high yam production costs.

CHAPTER 2

Yam Production Contexts in Nigeria and Ghana

T his chapter is based on information generated by the YIIFSWA baseline survey team in 2013. YIIFSWA is a five-year R&D project from 2011 to 2016 funded by the BMGF and implemented in Nigeria and Ghana by the IITA, Nigerian and Ghanaian national agricultural research institutions, and some nongovernmental organizations (NGOs). The goal of the project is to improve resource productivity in the yam food sector in West Africa by a 40 percent increase in yield through an improved seed system.

The YIIFSWA baseline survey was conducted to generate baseline data against which the project performance will be evaluated over time. Nigeria and Ghana account for 80 percent of West Africa's yam supply. In both countries, all yam agroecologies—humid forest, derived savanna, and southern Guinea savanna— were covered in the survey. In each agroecology, five communities were selected randomly in Nigeria and three in Ghana making a total of twenty-four communities, fifteen in Nigeria and nine in Ghana. In each community a stratified, random sample of three households was selected, totaling seventy-two, forty-five of which were in Nigeria and twenty-seven in Ghana.[1]

Data were collected through oral interviews of the selected farmers and through direct measurements of field size and yield. Field size measurements were done with a GPS; yield measurements were based on a sample plot of about fifty square meters that was harvested close to the center of the field. The survey was conducted in November and December when most of the mature yam was still in the field, unharvested. The YIIFSWA baseline survey had a narrow objective—namely to generate baseline data for project performance evaluation; therefore the information does not represent national averages because of the small sample size. The yam production contexts are discussed at village, household, and field levels.

Village-Level Conditions

The appalling economic conditions of yam-producing villages in Nigeria and Ghana are on display in the low levels of commercial activities, poor access roads, and shocking sources of water supplies in the villages. Village-level information collected in the YIIFSWA baseline survey shows that there are periodic markets in 20 percent of the Nigerian villages and 10 percent of Ghanaian villages. This means that there are no village markets in 80 percent of Nigerian yam-producing villages surveyed, nor in 90 percent of Ghanaian yam-producing villages surveyed. Availability of market centers in villages is a proxy for the level of commercialization; it is also presumed that villages without periodic markets are more remote from urban centers than others with such markets.[2]

The YIIFSWA baseline data further revealed that 25 percent of the villages surveyed in Nigeria and 10 percent in Ghana had all-weather motor vehicle access roads, while access roads to the other 75 percent of the surveyed villages in Nigeria and the other 90 percent in Ghana are hardly motorable during the rainy season. These observations of the YIIFSWA baseline survey show clearly that in Nigeria, and especially in Ghana, yams, which are sold in urban centers to urban consumers, are produced mostly in remote villages with low levels of commercial activity.

In Ghana in particular, people in virtually all the surveyed villages drew drinking, cooking, and washing water from puddles and rivulets that become stagnant or dried up during the dry season. The village people dig shallow wells on the beds of the puddles and rivulets when the water sources dry up completely. These dismal scenarios paint a graphic illustration of the high level of deprivation under which yam is produced in West Africa.

Yam-Producing Household Conditions

The number of years spent in formal education institutions by the heads of the yam farm households is low. In Nigeria, 27 percent had no formal education, and the situation is even more dismal in Ghana, where more than 70 percent of the heads of the surveyed households had no formal education (figure 1). Yam producers interviewed in the YIIFSWA baseline survey that had ten or more years of formal education in both Nigeria and Ghana were people that retired from urban wage employment and were only part-time yam farmers.

The low or zero level of formal education among yam farming households in Ghana and Nigeria should be of primary concern in R&D circles because the situation has far-reaching implications for efforts aimed at promoting yam production as

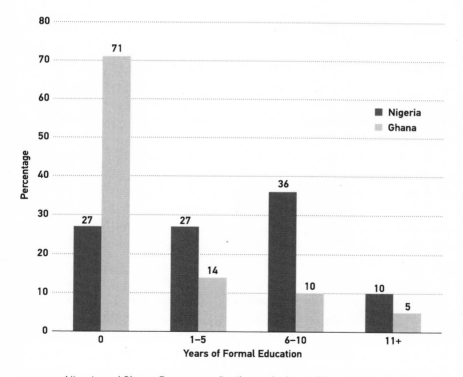

FIGURE 1. Nigeria and Ghana: Percentage distribution by level of formal education (years) of heads of yam-producing households, 2013. SOURCE: DJANA MIGNOUNA, ADEBAYO AKINOLA, ISSACQ SULEMAN, AND FELIX NWEKE, 2014, "YAM: A CASH CROP IN WEST AFRICA," YIIFSWA WORKING PAPER NO. 3, IITA, IBADAN.

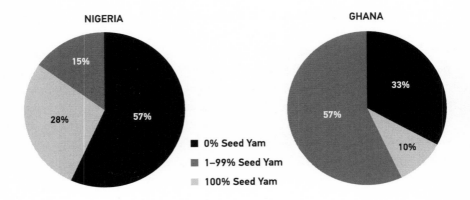

FIGURE 2. Nigeria and Ghana: Distribution of yam fields by percentage of seed yam purchased by country, 2013. SOURCE: DJANA MIGNOUNA, ADEBAYO AKINOLA, ISSACQ SULEMAN, AND FELIX NWEKE, 2014. "YAM: A CASH CROP IN WEST AFRICA," YIIFSWA WORKING PAPER NO. 3. IITA, IBADAN.

a business in West Africa in the twenty-first century. For example, yam production practices in both Nigeria and Ghana involve superstition and ritualism. In yam production, seed yam is the highest item of cost because the tuber, that is, the crop, is used as seed. The YIIFSWA baseline survey revealed that in both Nigeria and Ghana an average of 30 percent of yam harvest was used as seed, 30 percent for home consumption, and 40 percent for sale in Africa.

In Nigeria most farmers (57 percent) do not buy seed yam at all, and in Ghana most (also 57 percent) buy just a fraction of what they need even though it has been shown that the practice of recycling their own seed yam perpetuates seedborne diseases (figure 2). In Nigeria, Dr. C. C. Okonkwo, a former international yam trials manager at IITA, reported that one reason a farmer would not sell or buy seed yam was that selling seed yam could mean selling one's good luck and buying seed yam could mean buying someone else's bad luck.[3]

This farmer attitude toward buying and selling seed yam, which is seen as superstition, has a rational origin: seed yams available in the market are not quality declared, and buying poor-quality seed yams, which can result in poor crop performance, can be seen as buying bad luck. On the other hand, selling high-quality seed yam can rightly be interpreted as selling good luck because the high-quality seed yams will lead to good performance of the crop that could be envied, especially if the seller and the buyer are in the same vicinity.[4]

This negative attitude of farmers toward buying and selling seed yam has serious implications for efforts toward improving the yam seed system. Improved yam seed system is market driven and depends on farmer ability to pay and willingness to buy seed yam. The observed negative attitude means that some farmers who have the ability to pay for seed yam may lack the willingness to buy for fear of bad luck. Similarly the observed negative attitude means that there are farmers who would not produce seed yam for sale because of the fear of selling good luck to other farmers. Enhanced formal education can help a farmer to understand that another farmer who buys his seed yam deserves a good crop rather than envy. Elsewhere the need for farmer adult education in Africa has been emphasized (Nweke, Akoroda, and Lynam 2011).

Issahaq Suleman, a government extension officer in the Ejura district of Ghana, reported that farmers in that country do not apply inorganic fertilizers to yam because the farmers suspect that such fertilizers produce negative effects on the quality of yam.[5] Mr. Suleman further reported that in the same country, ritual objects prepared in clay or calabash pots were commonly sprinkled on seed yam before planting. After planting, the pot was left in the field to protect the yield of the crop from enemies because the farmers believed that, through ritualism, a farmer could transfer a good crop of yam in another man's field to his own. In fact, during the YIIFSWA baseline survey field work, pots filled with objects unknown to the YIIFSWA investigators were observed in yam fields visited in the Brong-Ahafo and Ashanti regions of Ghana.

The problem of superstitious and ritual practices in yam production has implications for the management of yam production as a business. The ritual practices present farmers with a multitude of issues. Crop failures are blamed on the enemy next door, and solutions to poorly performing crops because of pests, diseases, poor soil, or bad weather are sought in ritualism. Farmers who engage in superstitious and ritual practices are unlikely to be open to new technologies because the new technologies would be viewed with suspicion. Enhanced formal education among yam producers would help address the problem of superstitious and ritual practices that constitute an impediment to change in yam production in Nigeria and Ghana. The observed superstitious and ritual practices are clear evidence that there are human factors that must be considered in soil fertility and pest- and disease-management programs in yam production in West Africa.

One reason for limited R&D attention to laborsaving technologies in African agriculture is the wrong assumption that, relative to other inputs such as inorganic

TABLE 2. Nigeria and Ghana: Household size (no./household), 2013

STATISTICS	NIGERIA	GHANA
No. of Households	41	21
	NO./HOUSEHOLD	
Mean	11.54	11.95
Min.	1	3
Max.	35	25
Std. Dev.	8.3099	6.0620

Source: Data from Djana Mignouna, Adebayo Akinola, Issacq Suleman, and Felix Nweke, 2014, "Yam: A Cash Crop in West Africa," YIIFSWA Working Paper No. 3, IITA, Ibadan.

fertilizers, farmers have enough labor because of the large farm household sizes.[6] The large yam farm households that were observed in the YIIFSWA baseline survey have an average of about twelve persons per household with the range of one to thirty-five, in both Nigeria and Ghana. This gives a false impression of a high level of availability of farm labor from household sources (table 2). Farm household size as a proxy for labor availability depends on the composition of the household. Many of the large households encountered in the YIIFSWA baseline survey were composed of aged women in polygamous families and many children either school-aged or younger, whose contributions to farm work would be low or nonexistent. This means that household size can be a misleading proxy for labor availability in yam production.

Conditions at the Yam Field Level

The YIIFSWA baseline survey data show that most of the yam farm households had one yam field each; in both Nigeria and Ghana, less than 5 percent had more than one. Average yam field size was 1.82 hectare (ha) per household in Nigeria and 1.60 ha in Ghana. Average yield was approximately the same, twenty-six tons per ha in both countries (tables 3 and 4).

In Ghana the yam fields surveyed were situated at distances of up to fifteen kilometers from the village centers. In that country, yam is produced under the shifting cultivation system; each season famers go into forests in search of suitable

TABLE 3. Nigeria and Ghana: Yam field size (ha/household) by country, 2013

STATISTICS	NIGERIA	GHANA
No. of yam fields	39	21
	FIELD SIZE/HOUSEHOLD	
Mean	1.82	1.60
Min.	0.22	0.11
Max.	12.32	6.74
Std. Dev.	2.2543	1.7400

Source: Data from Djana Mignouna, Adebayo Akinola, Issacq Suleman, and Felix Nweke, 2014, "Yam: A Cash Crop in West Africa,"YIIFSWA Working Paper No. 3, IITA, Ibadan.

TABLE 4. Nigeria and Ghana: Yam yield (tons/ha) by country, 2013

STATISTICS	NIGERIA	GHANA
No. of yam fields	40	20
	TONS/HA	
Mean	26.03	25.65
Min.	5.29	9.89
Max.	81.81	71.81
Std. Dev.	17,383.60	17,502.58

Source: Data from Djana Mignouna, Adebayo Akinola, Issacq Suleman, and Felix Nweke, 2014, "Yam: A Cash Crop in West Africa,"YIIFSWA Working Paper No. 3, IITA, Ibadan.

land for yam production without ever returning to land that had previously been farmed.[7] The result is long and increasing distances between farmers' homes and yam fields. The fields are along forest tracks with thick bushes of sharp grasses such as *Imperata* and across rivulets, some of which are knee-deep. On-farm transportation is on foot, by bicycle, or by motorcycle for men and on foot for virtually all women. The women have to leave home early in the mornings and return late in the evenings with loads, carried on their heads, of firewood in the planting seasons and firewood and crops in the harvesting seasons.

In Nigeria, the context is dismal but less so than in Ghana because in Nigeria yam is produced under long fallow rather than under shifting cultivation. In Nigeria, farmers return to grow yam on land previously planted with the crop every three to five years. During the years in between, other crops are grown on the land.

Distances between the farmers' homes and yam fields are not as long as in Ghana, nor are they increasing.

A Possible Cause of Poverty among Yam Producers

Yam is produced more for sale than for home consumption; in both Nigeria and Ghana 60 percent of the harvest, after discounting for seed, is sold and only 40 percent is consumed in the farmers' households.[8] The crop attracts a high price in the urban markets because it is patronized by high-income consumers. Producers work hard to produce so much yam, yet they live in penury. Why? Two related explanations are relevant.

First, low postharvest technologies in the yam food sector render producers helpless at the price negotiating table with the yam traders. Yam is bulky, and though postharvest shelf life can be up to six months depending on variety, during the storage period losses come from several sources including weight loss from dehydration, sprout growth, and more importantly from a wide range of pest and disease attacks, especially attacks by nematodes and viruses.

Second, there is a problem at the level of outlet for yam from the farm to urban market. In Ghana, a yam producer faces three options when disposing of his crop:

1. The producer takes a loan from a trader at planting time to pay for the high production costs and at harvest the trader arrives at the farm with a truck and carries the yam away at his price in return for the loan.
2. A trader arrives at the farm of a producer who did not receive a loan and makes a price offer; if the price is not acceptable to the producer the trader goes to another farm with his empty truck.
3. The farmer takes a truckload of his yam directly to the urban market and stops at the entrance. A league of middlemen prohibits the farmer from entering the wholesale market with his yam. Right at the entrance to the market, a middleman takes the truckload of yams from the farmer and negotiates the price with wholesalers behind the farmer's back and takes his agreed and unagreed commission before handing the proceeds over to the farmer who does not know what the wholesaler paid.[9] If the farmer does not like the amount he is offered, he can take his truckload of yam back to the farm, but this is an expensive option for him.

Without question, the context in which yam producers sell their crop has a high potential to impoverish them; policy interventions are needed to change the unfair situation. The first step is to empirically assess the marketing situation to determine if the yam traders are enriched by the situation that impoverishes the farmers. The empirical assessment will identify measures that, if implemented, will enable all participants in the yam value chain, the producers as well as the traders, to be equitably compensated for their efforts.

Making effective changes that will move benefits from one participant in the value chain to another in an established marketing system with entrenched interest groups is a serious agenda because it is highly political. To succeed, changes must be approached through advocacy among political leaders at various levels of governance and among business union groups. Obviously such advocacy must be supported with concrete information on who gains and who loses under the status quo and how these people will be affected under proposed changes in the status quo. This information can be obtained through in-depth marketing R&D in order to be convincing to political leaders, policy makers, researchers, and the wide range of participants in the yam value chain.

Summary

In Nigeria and Ghana yam is mostly produced in villages that are remote and with poor road access to market centers. Low quality of life in the yam-producing villages, especially in Ghana, is on graphic display in the dependence of the village people on puddles and rivulets, which are stagnant in dry seasons, for drinking, cooking, and washing. Little or no formal education predisposes the mainline farmers in both Nigeria and Ghana to unorthodox yam production practices, such as reliance on rituals in preference to trading seed yam.

Yam-producing households are large in size, an average of twelve persons per household in both Nigeria and Ghana, but that does not translate into availability of labor as hired labor is more important than family labor in yam production. Several of the household members are either aged men and women or school-aged children who are physically incapable of executing most yam production tasks.

Among the most critical constraints in yam production in West Africa is the practice of shifting cultivation, which mostly happens in Ghana. The practice exposes the farmers in that country to unproductive and tortuous commuting between their homes and their yam fields because the practice entails continuous penetration of the forest that creates increasing distances between the yam fields and farm homesteads. The practice of shifting cultivation, which is rooted in the farmers' continuous search for fertile land, low yam pest and disease incidences, and stake trees, has negative implications for environmental degradation.

The yam marketing system works against the farmers, forcing them to live in penury in spite of their backbreaking yam production engagement. Determinations of the obstacles against farmers receiving compensation commensurate to their efforts and identifying and implementing policy measures to effectively eliminate those obstacles are essential in order to bring down the high level of poverty in yam-producing communities in West Africa. Such changes will require advocacy among political leaders at various levels of governance and among business union groups.

CHAPTER 3

High Labor Cost in Yam Production

igh labor input is one of three difficulties that stand in the way of yam production expansion and cost reduction in West Africa; others are low seed
technology and pest and disease problems. In West Africa, all food crops
are produced at high labor costs because of dependence on Iron Age laborsaving
technology; certain agronomic practices that are peculiar to yam production and
have defied change with time add up to make yam production labor intensive in
the extreme.

Yam and Cassava Production Labor Inputs in Nigeria

On a per unit area basis, aggregate yam production labor is one and a half times
more than cassava (table 5). Why? The hand tools are the same: a hand hoe and a
machete, which are made by blacksmiths and carvers in the farmers' villages. The
blacksmith molds the blade, and the carver carves the wooden handle of the hoe
or the machete. But agronomic practices are different between the two crops; yam
is grown in mounds in virtually all cases, while cassava is grown on flat or ridged

TABLE 5. Labor inputs in yam and cassava productions, person days/ha

FARM TASK	YAMS	CASSAVA
Land Preparation	51	24
Planting	33	29
Staking	59	0
Weeding	45	31
Harvesting	26	52
Total	214	136

Sources: The yam data comes from Nweke et al., "Production costs in the yambased cropping systems of southeastern Nigeria," RCMP Research Monograph No. 6. (Resource and Crop Management Program, IITA, 1991); the cassava data comes from Tshiunza, "Comparative study of production labor for selected tropical food crops (cassava, yam, maize and upland rice)," in *Root crops and poverty alleviation*, ed. M. O. Akoroda and I. J. Ekanayake, 121–25. Proceedings of the Sixth Triennial Symposium of the International Society for Tropical Root Crops–Africa Branch, Lilongwe, Malawi, October 22–28, 1995.

seedbed most of the time. In fact, cassava is grown on mounds mostly when it is an intercrop with yam.

Mound-making labor, fifty-one person days per ha, is considerably higher than plowing and ridging labor, twenty-four person days per ha (Tshiunza 1995).[1] Yam is planted after fallow more frequently than cassava, and cassava is planted before fallow more frequently than yam. The labor for clearing fallow fields and the high labor for mound making combine to account for the considerably higher yam farmland preparation labor than cassava.

Cassava planting setts are bulkier than yam planting setts, yet yam planting is marginally more labor intensive than cassava. Seed yam is handled with care because it is a high value item. Additionally, mulching, which is not relevant to cassava, is considered part of the planting task because planting and mulching are done at the same time.

Staking, which is peculiar to yam because the plant is a climber, is labor intensive. In a growing season, staking of yam is not a once and for all activity. From time to time, the farmer goes back to trail the vine branches up the stake; otherwise the vines will grow in the wrong direction and could even crawl on the ground, negating the effort of staking. In most yam-growing areas of Nigeria, staking materials used by the farmers are bamboo branches, or splits, or shrub branches for small size yam categories such as seed yam, and full bamboo stems or similar wood materials for large yam size categories such as ceremonial yams. Most of these materials do not last more than one crop season because they rot in the wet, humid environments and become targets of termite attacks. Wooden stake materials, when recycled, could be an alternate host to some of the wide range of the yam pests and diseases.

In most yam-growing areas in Ghana, the staking practice is different from the practice in Nigeria; in Ghana yam is planted in fields where stakes are preexistent. How does this work? Each planting season farmers move to virgin forests; they slash and burn the forest in order to clear land and then collect the debris, which is burned around small trees that die and serve as stakes for the yams. This destructive practice is extremely labor intensive; it forces farmers to move deeper into forest lands one season after another in search of sites that contain suitable stake trees, thereby constantly creating increasing distances that they have to travel on a daily basis between village settlements and yam fields.[2]

In the 2013 baseline survey of yam production conducted in Ghana in the context of the IITA's YIFFSWA R&D project, distances of up to ten to fifteen kilometers between village settlements and several yam fields were measured. Commuting between the settlements and the yam fields by wading through thickets of bush and crossing rivulets, some knee-deep, over such distances on bush tracks on a daily basis is time consuming and fatiguing. This unproductive activity is a torture for both men and women, but more so for women. While some men are able to ride bicycles or motorcycles to the fields, women commute on foot leaving home early each morning and returning late in the evening with loads of firewood carried on their heads during planting season and loads of firewood and crop carried on their heads during harvesting season.

Weeding yam fields is substantially more labor intensive than weeding cassava fields for a number of reasons. The nuisance of weed is higher in yam production than in the production of cassava because yam is more frequently grown after fallow and cassava more frequently before fallow. Both yam and cassava have long growing seasons compared to most other food staples in West Africa, but in a crop cycle, yam is weeded a larger number of times than cassava. Yam is weeded frequently close to harvest while cassava, a hardier crop, is left in the field, in many cases, beyond full bulking period as a storage method, without further weeding. Additionally, performing any farm task is more laborious in a mound field where the mounds are set haphazardly than in fields of ridges, because ridges have some measure of order.

The last of the yam production field tasks, namely harvesting, is carried out with digging sticks, machetes, or small hand hoes at the end of the growing season, usually after the soil has dried up, making the harvesting operation a more arduous task. When farmers need to hire labor to harvest their yams, they look for experienced hands because wounds to the tuber at harvest are easily translated to waste in storage because such wounds provide easy access for various yam storage pests and diseases. The only way yam is harvested is by digging, sometimes deep when yield is high because the tubers grow vertically unlike cassava, which can be harvested by lifting, especially when the soil is wet because the roots grow laterally. Yet yam harvesting labor is considerably lower than cassava because yam is harvested in relatively weed-clean fields while cassava is often harvested from fallow fields.

Cassava harvesting labor presented in table 5 is unusually high because of high yields obtained by planting IITA's high-yielding mosaic resistant varieties in Nigeria. Cassava harvesting labor per unit area increases in direct proportion with yield because of bulk; cassava is 70 percent water compared with white and yellow yams, which have an average of 30 percent water.

Lack of Technological Change in Yam Agronomic Practices

During the yam consumption survey, farmer groups, which included the oldest farmer in the village, were interviewed in Nigeria and Ghana and were asked to list the five most common seedbed types they use for yam and about what year each was introduced to their area. In all groups only mound was listed, and each farmer group reported that they did not know when planting yam in mounds started. Yam researchers in West Africa are now challenged to fathom the underlying reason for the farmers' dogmatic reliance on mounds for yam production against available scientific evidence of superior laborsaving and other advantages of ridges.

In eastern Nigeria ceremonial yams are planted early in the season before the rain regularizes. They are mulched with leaves to conserve moisture and protect the young sprout, when produced, from the direct rays of the sun. This works well except that the mulch hosts an array of the yam plant pests and diseases. Yam scientists in West Africa have shown that plastic sheet is superior to leaves as yam mulch material because the plastic sheet sheds off weeds, controls soil temperature, conserves moisture better, and does not host yam pests and diseases. In fact, scientific experiments established that yield is higher in plastic-sheet-mulched yam than in leaf-mulched yam.[3] In spite of the scientific evidence in favor of plastic mulch, farmers continue to use leaves as mulching materials. This is probably because of the convenience of availability and lower cost. The plastic mulch, which is not biodegradable, is difficult to remove after the season, constituting extra cost.

Staking has provoked much interest among yam researchers aiming to establish the effect of the practice on yam disease and pest control, weed suppression, and yield, essentially because the activity is an expensive agronomic practice.[4] Staking was reported to produce yield advantage over not staking (Ndegwe 1992). Input-output analyses to establish if the yield advantage sufficiently compensates for high staking costs are called for. Yam staking is expensive because it is labor intensive and the stakes, whether purchased or collected by the farmer, have private and social costs. As yam production expands, demand for wood for stakes increases, exerting pressure on the environment. Perhaps one day an entrepreneur will see the business opportunity in supplying the growing demand through yam stake farming. There is already a growing market for yam stakes in parts of Nigeria, but supplies are collections from the wild.

Yam researchers in West Africa have experimented with alternative staking materials and found that PVC pipes and plastic mesh proved to be suitable alternatives to wood because they are not damaged by termites, they can be reused for a number of growing seasons, and they may not play host for yam pests and diseases when recycled (IITA 1974). Farmers dogmatically stick to using wooden stakes. The farmer groups interviewed in the three major yam-growing centers in Nigeria and two in Ghana during the yam consumption study all reported that wooden stakes were the only material they used and that they did not know when sticks were first used as yam staking material. The underlying reason for this practice by farmers in continuing to use wood for yam staking materials when research has identified alternative materials with several advantages is an empirical question.

In summary, Iron Age laborsaving technologies are employed in the production of yam and other food staples such as cassava. Yam production is more labor intensive than cassava production because of yam's peculiar agronomic practices such as planting in fertile land after fallow, planting in mounds, mulching, staking, and high weeding frequency.

Yam Production Labor Profile

High aggregate labor input in yam production is one thing; skewed distribution with high peak demand periods is another. Available information on biweekly yam

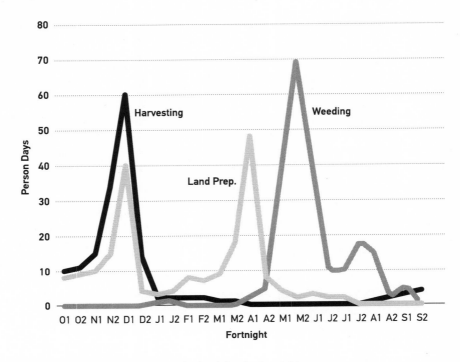

FIGURE 3. Eastern Nigeria: Yam production labor by fortnight. source: ugwu 1990.

labor distribution for the derived savanna agroecology of eastern Nigeria reveals that yam production labor demands for various tasks are characterized by steep seasonal peaks (figure 3). Land preparation was from mid-October to April with peaks in November and December and in March and April. In the area where the labor information was obtained in eastern Nigeria, the soil was lateritic and difficult to till, both in the dry season when the soil was hard and in the wet season when the soil tended to form clods. Consequently, land preparation was done between the dry and the wet periods. A minimal level of land preparation was done in September for ceremonial yam in hydromorphic soil.

Planting of yam started slowly in December, peaked in early April and late May, and declined to zero in July. This distribution is because yams of different size categories are planted at different times. Some of the early planting was ceremonial yam in the hydromorphic land to maximize available insolation, which became

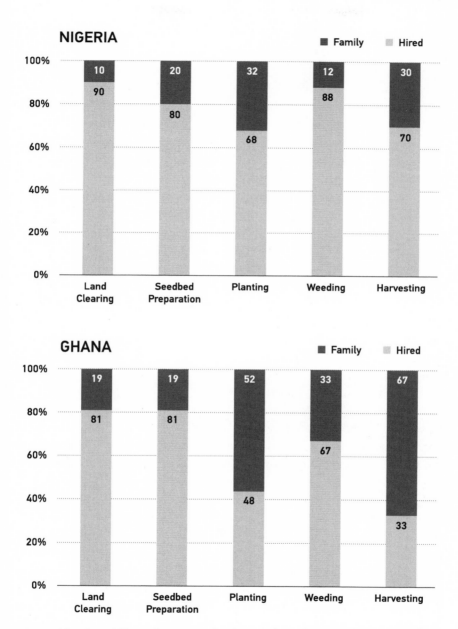

FIGURE 4. Nigeria and Ghana: Percentage distribution of yam fields by source of labor by farm operation, 2013. SOURCE: DJANA MIGNOUNA, ADEBAYO AKINOLA, ISSACQ SULEMAN, AND FELIX NWEKE, 2014, "YAM: A CASH CROP IN WEST AFRICA," YIIFSWA WORKING PAPER NO. 3, IITA, IBADAN.

limited later during the growing season. Late planting for the production of seed yams could be as late as July.

Staking labor is at a low level over long periods of time from January until July. Early staking is for ceremonial yams, which are planted early in the season, while later staking with peaks in labor demand are a combination of two factors: initial staking of seed yams, and trailing of vines of ceremonial and table yams that were staked earlier. Harvesting labor input starts around July and August for ceremonial yam, which is planted early; the harvesting labor demand peaks in November and December. Weeding labor demand begins in March or April for the early planted yam fields and peaks around June. The second peak of the weeding labor demand, which is low compared to the other peaks, is around July and August for late-planted yam fields, usually seed yam fields.

The YIIFSWA baseline survey data reveal that yam production uses more hired labor than family labor in both Nigeria and Ghana (figure 4). Farmers rely on seasonal migrant laborers who come from outside of the area. Everyone in the area is usually busy with their own work. Managing the hired labor is an issue; paid on a daily basis, the migrant laborers spent a considerable amount of time idling, waiting for closing time. When they were paid by individual task, they rushed through their work to save time, and their performance was shoddy unless closely supervised. So the farmer had to stay with one labor team and was not able to engage hired laborers in more than one field at a time.

High and increasing farm wage rates are yet another serious issue the yam farmer contends with. In the Otuocha area, during the land preparation season of 2013, the daily wage rate was equivalent to US$12 per person per day, up from US$10 in the 2011 land preparation season.

Cost Profile in Yam Production

In the forest ecology of eastern Nigeria, the cost of seed yam was at least 50 percent of the total cost of yam production (Nweke et al. 1991). The proportion was so high because the tuber, the edible part of the yam, was used as planting material. Tuber used as seed yam amounted to 30 percent by weight of the total output in the forest ecology partly because of the production of ceremonial yams. Seed yam used to produce ceremonial yam is the largest in size among the seed yam tuber size categories. Labor cost is not far behind, at more than 40 percent of the total yam production cost. Contributions of other items, such as stakes, fertilizer, and pesticides, to the total cost of yam production were low in percentage terms.

Insights

These findings point to high potentials in mechanical laborsaving technologies for cost reduction in yam production. Reduced production costs will drive down the price of yam to consumers, expand demand for yams, and raise producers' incomes through expanded production and sales. Extensive literature research yielded limited, if any, evidence of completed or ongoing efforts in the development of mechanical laborsaving technology for any of the yam production tasks. Reliance on imported prototype machines from Europe, North America, or Asia that have not worked for other crops in the past has a lower chance of success in the case of yam, which is a West African crop that is better understood by design engineers in the region.

Designing laborsaving mechanical technology for yam production and harvesting calls for imaginative thinking because of characteristic agronomic practices such as mound making, staking, and harvesting by digging vertically, which are peculiar to yam. It calls for imaginative thinking also in terms of the nature of the machinery. Improved hand tools that are compatible with African agriculture systems consisting of low capital base smallholders who cultivate small fragmented plots, fragile soils held together by shrub stumps, etc. would be more relevant than tractor mechanization. Such design requires tripartite effort among design engineers, scientists, and farmers in West Africa who understand the yam culture.

CHAPTER 4

Yam Seed Technologies

S eed yam accounts for more than 50 percent of yam production cost in eastern Nigeria, and up to 30 percent of the crop produced is used as seed in the subsequent planting season.

Sources of High Cost of Seed Yam

Yam is propagated by tuber, the edible part of the crop, with a very low multiplication rate. A seed yam is a small whole tuber or cut pieces of a larger tuber. The low multiplication rate is made worse by the long growth cycle and long dormancy of yam. Yam growth period is five to nine months depending on variety and the yam tuber size category; the dormancy period is two to three months after senescence. The worst scenario, which is common in practice, is that when a small whole tuber is used as seed yam one plant can produce one seed yam after one year, especially if the yield is another small size tuber, which is sometimes the case. At best, if one tuber produces five seed yams through planting cut pieces of it, after one year the original tuber could generate five seed yams, that is, before discounting for losses due to yam pests and diseases. This is low compared with as much as two hundred seeds that can be obtained from one plant of grain in a three-month period. This low rate of multiplication and use of the tuber—the edible part—for propagation makes seed yam expensive in the extreme.

As a clone, seed yam carries over seedborne pests and diseases from one generation to another. Seed yam is prone to the situation in which yields of crops decline because of the progressive build-up of pests and diseases in the propagation stock following systemic infestations and infections transferred through each generation (Gibson et al. 2009). The seed yam–borne pests and diseases affect seed productivity and viability, reducing germination, plant vigor, and yield (Aighewi 1998; Ezeh 1998). Yam production with diseased seed yams results in small, poor-quality table yam and a persistent cyclical decline in yield. These scenarios mean that the cost of producing seed yam is more than the price paid for it in the market because of the loss in yield, which results from low quality. Planting of healthy seed yams is therefore central to improved productivity in yam production, but access to healthy seed yam is one of the biggest problems for the yam producers (Aidoo et al. 2011; Coyne, Claudius-Cole, and Kikuno 2010).

Yam Seed System in Nigeria and Ghana

In Nigeria and Ghana, the yam seed system is predominantly informal; farmers function, without public regulation, to provide seeds for their own planting or for exchange, sometimes through the market mechanism. Since the informal system is not publicly regulated, the quality in terms of health and varietal purity of seed is not guaranteed (Nweke, Akoroda, and Lynam 2011). The present informal seed yam system is driven by market mechanism. For example, in Igalaland of Kogi State of Nigeria farmers buy seed yam each growing season rather than producing their own; they seek quality judged by physical inspection for symptoms of damage by pests and diseases.[1]

In Nigeria and Ghana, seed yam production methods, which vary as widely as yam crop production methods, can be classified into two broad groups. Separation of seed yam production from yam crop production is the central distinguishing practice of one broad group, which is practiced in Nigeria along the River Niger basin from around the confluence of the Niger and Benue Rivers in Lokoja down to the southern limit of the Nigerian yam belt. Seed yam is produced by planting extra small seed tubers or small pieces of regular sized tubers at later dates than yam crop planting and often in soils not fertile enough for yam crop production. This practice is often described, in yam research circles, as specialized seed yam production; the same farmers produce yam crop, although independently of seed yam production.

Production of seed yam as an integral part of the yam crop production is the second broad group. It is more widely practiced throughout the yam belts of Nigeria and Ghana than the first broad group described. Within the second broad group, two subgroups of practices are distinguishable: milking and sorting. Milking is practiced with early maturing yam varieties such as the *pona* local variety; close to senescence the crop is harvested while the plant continues to grow and regenerate a multitude of seed yams (Osei-Serpong undated). Undersized crop is often buried back in the yam mound seedbed, in what is described in yam research circles as coping, to be later used as seed yam. This practice represents significant investment by the farmer. First, there is a yield loss by harvesting the main tuber before senescence when maximum yield would be attained (Onwueme 1977). Second, premature harvest reduces the quality of the main tuber in terms of both reduced shelf-life and texture. Third, harvesting labor is increased because of the multiple harvests. These costs help explain the very high cost of seed yam.

In a yam crop field, sizes of tubers produced vary; depending on microenvironmental variations within the field, such as variations in soil fertility and in soilborne yam pests and diseases. In addition, occasionally a yam plant produces multiple tubers of varying sizes. At harvest, tubers are sorted by size: small ones for seed yam, medium ones for table yam, and the very large ones for ceremonial yam. Use of seed yams produced by sorting facilitates the recycling of yam pests and diseases, especially those seed yams that are small because of microenvironmental variations in the field.

The informal seed yam systems do not deliver quality seed yam and do not deliver a sufficient quantity of seed yams to the production system in both Nigeria and Ghana where the demand for seed yams exceeds supply virtually always and prices are high. Every planting season, farmers commonly use cassava to fill empty seedbeds in yam fields created by a shortage of seed yam.

The farmer groups interviewed in the yam consumption study in the major yam producing centers in Nigeria and Ghana were asked to name and rank their seed yam technologies in terms of material they use as seed yam. Only three materials were mentioned: whole tuber, regenerated tuber from milked yam plant, and cut up tuber pieces (table 6). When asked about what year the material they use as seed

TABLE 6. Nigeria and Ghana: Common seed yam technologies by popularity rankings (1 represents the most popular material used), December 2012

COUNTRY	YAM CENTER	YAM MATERIAL	RANK
Nigeria	Shaki	Whole tuber	2
		Regenerated tuber	1
		Sliced tuber	3
	Otuocha	Whole tuber	2
		Regenerated tuber	Not used
		Sliced tuber	1
	Zaki Ibiam	Whole tuber	1
		Regenerated tuber	Not used
		Sliced tuber	2
Ghana	Tamale	Whole tuber	2
		Regenerated tuber	1
		Sliced tuber	3
	Kintamkpo	Whole tuber	2
		Regenerated tuber	1
		Sliced tuber	3

Source: Data from Felix I. Nweke, Robert Aidoo, and Benjamin Okoye, 2013, "Yam consumption patterns in West Africa," unpublished report submitted to Bill and Melinda Gates Foundation.

was first used in their area, the farmer groups greeted the question with derisive laughter. That was not a good question; the investigators should have known that the material was used from the beginning of time.

In eastern Nigeria, material ranked number one in popularity was determined by the size of yam tuber being produced. Farmers who practice specialized seed yam production use extra small tubers or slices of larger tubers as seed yam for producing bigger seed yams, which is used to produce table and ceremonial sized yam tubers in the following season. Other farmers who produce ceremonial yams, such as River Niger flood plain yam farmers, use large whole tuber seed yams because there is a positive relationship between seed yam size and size of tuber harvested.

In Ghana and several yam-producing areas of Nigeria, regenerated tuber from milked yam plant was the number one material used for seed yam. In Ghana, preferred yam varieties, namely *pona* and *laribako*, are not efficient with the Nigerian farmers' technique of planting slices of table tuber. The preferred varieties do not have sufficient nodes (sprouting eyes) over the surface area of their tuber as necessary for the Nigerian farmers' technology. Farmers in Ghana are able to grow large ceremonial yams with their common seed yam technology because regenerated bunches of tubers from milked yam plant contain different sized tubers. Some of these tubers are big enough for ceremonial yam production.

Advances in Seed Yam Technology

The aim of improving seed yam technologies is to transform the yam seed system from an informal system to a formal system. In the formal seed system, quality-declared seed is generated in two ways. A chain of stakeholders beginning with breeders develop improved varieties of seed or farmer-researchers select, from existing stock, local cultivars with farmer-desired attributes and clean them of pests and diseases. The seed yam produced at this stage is described as breeder seed. The chain continues with stakeholders described as foundation seed producers, who take the breeder seed yams, usually available in small quantities, and multiply them until there is enough to go to the next link in the chain, specialized seed farmers. Specialized seed farmers multiply the foundation seed yams to sell to farmers in sufficient quantities at affordable prices; the chain goes on to include seed processors and distributors for packaging and marketing. The stakeholders at the various links in the seed chain function under the inspection of public seed regulatory agencies and under public seed law that has sanctions for noncompliance. Only seed certified and declared as quality seed by an authorized seed agency under the law can be offered to farmers.[2]

Researching literature on the subject of seed yam reveals that in the past thirty to forty years significant progress was made toward improving the efficiency of the traditional tuber seed yam technology by developing methods that can enhance the multiplication rate. More recently, considerable progress is being made toward developing alternatives to the tuber seed yam technology.[3]

One of the technologies that aim to improve on a traditional method by increasing the multiplication rate, that is, increasing the number of seed yams per plant, is known as the minisett technique in yam research circles. The minisett technology, which was developed at Nigeria's National Root Crops Research Institute (NRCRI) in the early 1970s, utilizes a small (20–50 gm) part of a whole nondormant tuber containing periderm and some cortex parenchyma (Okoli and Akoroda 1995).[4] By cutting up a tuber into many small pieces and subjecting the pieces to appropriate treatments to facilitate sprouting, several seed yams can be produced from one tuber, which under the traditional method can give only a few seed yams. The technique comes with a list of nine instructions including how to select the right tuber to cut, how to cut the tuber, chemical treatment, planting, etc. (Maduekwe, Ayichi, and Okoli 2000). Minisett technology has provoked more scientific investigations than any other subject in yam research in West Africa. The technology is investigated from different angles in order to provide feedback for improvement.

Several researchers studied the effects of alternative minisett sizes above the originally recommended 20–50 gm on sprouting and tuber yield concluding that total production cost per hectare, yield, and net return increased with increasing minisett size.[5] Some other researchers investigated alternative intercropping using minisett production with the aim of making the minisett technique acceptable to the smallholder farmers who practice intercropping.[6] One such intercrop study reported a minisett yield decline of about 33 percent using an intercropping system (Ikeorgu, Anioke, and Nwauzor 1998). Other investigators studied the response of different yam varieties to the technique. In Ghana, studies that investigated the response of a popular white yam variety reported that the *pona* variety showed a poor response to the technique (Yankey 2002). There were studies of alternative chemical treatments as well.[7] The study that investigated the response of the *pona* to minisett technology in Ghana reported that treatment with naphthalene acetic acid improved response.

Economic profitability studies return mixed results; production of seed yam with minisett technology breaks even under certain conditions including elevated seed yam prices.[8] Numerous adoption studies yield discouraging results of low adoption and in some cases reverse adoption.[9] Farmers continue to practice their seed yam production using traditional methods without the improvements that minisett technology offers. Why?

The complexity of applying the technique, the requirement of chemical treatments, and high labor inputs are implicated in the low adoption by the various adoption studies. Yet if the technology worked, farmers would invest in learning the technique, in buying chemicals, and in hiring labor. More important than those challenges are the effects on adoption of both poor sprouting and the positive relationship between sett size and yield. This means that for the technology to work, the causes of poor sprouting and of the need for larger setts than recommended should be eliminated. Cut pieces of yam are more susceptible to nematodes, viruses, and fungi than whole tuber seed yams; the smaller the setts, the greater the difference in susceptibility. The implication is that finding solutions to those yam pests and diseases is a precondition for the diffusion of the minisett technique.

The alternatives to tuber seed yam that are at various stages of development in West Africa by yam researchers include producing and germinating true seed of some yam varieties, rooting yam peels discarded at yam food preparation, rooting stem cuttings of yam plant, and rooting yam sprouts generated by yam in storage after dormancy. Yam peels produced during yam food preparation and sprouts produced by yam in storage are discarded because there is a lack of use for them. Yam plants grow enough vegetation that taking cuttings from them would not result in a noticeable cost in yam production. Therefore, developing viable alternatives to yam tuber seed from these materials would help reduce seed yam production costs, drive down the price of yam to consumers, and lead to increases in producer income.

Yam peels, sprouts, and vine cuttings have been successfully rooted and used to produce tubers in greenhouses with soil medium, in a laboratory with tissue culture, and with the aeroponics technique (Maroya et al. 2014). In most cases, the tubers produced by these techniques under the greenhouse conditions would require one or more growing cycles to attain the correct tuber size for producing seed yam. Yet the developments of these technologies qualify as breakthroughs and represent considerable achievements.

At this stage, private entrepreneurs that can provide necessary logistics, such as laboratories and greenhouses, should take over, advance the technologies, produce seed yams of required sizes and quality, and sell to farmers. The experience of the prevailing informal system shows that a market exists for seed yams when made available at the right time and at affordable prices, but the present macroeconomic policy environment in major yam-producing countries in West Africa may not favor private investment in enterprises such as seed yam production that demand intellectual property protection and reliable power and water supplies. In addition, because of the small sizes of most of the first-generation tubers produced, the techniques face the challenge of absence of precondition for adoption, namely control of yam nematodes, viruses, and fungi.

Summary

The high cost of seed yam is a major bottleneck in yam production because the edible tuber, that is, the crop, is used as seed. Up to 30 percent of annual production is replanted as seed, which increases seed yamborne pests and diseases and decreases yield. Yam researchers in West Africa are making steady progress toward developing novel seed yam technologies that can overcome the problem of using expensive, low-quality tubers as seed yam. The greatest challenge is to provide the necessary precondition for adoption, namely the control of yam pests and diseases, especially nematodes and viruses.

CHAPTER 5

The Challenge of Nematodes in the Yam Food Sector

S everal pests, viruses, and fungi constitute serious problems along the yam value chain from seed yam production through yam crop production, storage, marketing, and consumption. Among all of the pests and diseases, the nematode problem could be considered the first bottleneck because, in most cases, the nematodes create avenues for attack by several yam diseases and other pests (Coyne, Nicol, and Claudius-Cole 2007; Coyne, Claudius-Cole, and Kikuno 2010; Nwauzor 1998b).

The Nematodes of Yam

Nematodes are tiny worms; they are ubiquitous, occupying almost every agricultural niche, but favor moist environments rather than dry (Asante, Wahaga, and Ramat 2002). Three nematodes that cause the most problems in yam are known as yam nematode, root-knot nematode, and lesion nematode (Nwauzor 1998b).

Yam nematode invades the yam plant at the primary root growing point, through cracks, and through damaged areas on the surface of the tuber while it is in storage. Yam nematode inhabits the periderm and subperiderm layers of the tuber, but it is migratory and while moving in the yam tuber tissue it disrupts and empties the yam cells. Small yellow necrotic lesions develop under the skin of the tuber during early infection, which later turn dark brown and eventually coalesce to form a continuous layer beneath the surface, a condition known as dry rot (Coyne, Nicol, and Claudius-Cole 2007; Coyne, Claudius-Cole, and Kikuno 2010; Nwauzor 1998b).

Root-knot nematode penetrates the roots and tubers of yam through the softer meristem portions, causing hyperplasia and hypertrophy of cells. It induces galls on the roots or on the tuber, giving it a warty appearance.[1] In some cases, root-knot nematode causes proliferation of rootlets on the tubers. Lesion nematode penetrates tubers causing necrosis that leads to dry rot of the periderm and subperiderm regions similar to those caused by the yam nematode (Nwauzor 1998b).

The skin of a nematode-affected yam tuber develops cracks that admit viruses, bacteria, fungi, etc. into the tuber. The resultant complex gives rise to wet rot of internal tissues, destroying the entire tuber. Nematodes survive in the tubers through the storage period, and the tubers serve as a source of inoculums for spreading. Seed yam serves as a vehicle for affecting yam in the field and harvested yams as a vehicle for affecting yams in storage (Nwauzor 1998b).

Nematodes cause yield reduction and loss of the edible portion in stored tuber. With destruction of the meristem zone, sprouting of the affected tuber is impaired. In the field, nematode-affected seed yams sprout, establish, and perform poorly. Nematodes survive in the soil and debris, while host plants include many cultivated and uncultivated crops (Krapa 2006; Nwauzor 1998b).

Presently cultivated yam varieties are susceptible to the nematodes, but the nematodes can be controlled in yam by using nematode-free seed yam; thermo- or chemotherapies can be used to ensure planting of nematode-free seed yams. Soil exposure and tillage at the peak of the dry season to expose the nematodes to the harsh weather help to bring down nematode soil population. Rotation and long fallow that do not include host crops help to reduce nematode population in the soil as well (Nwauzor 1998b).

Costs of Nematode Problem in the Yam Crop Sector

Nematodes cause damage in terms of crop losses to yam both in the field and in storage, but the literature offers few, if any, objectively measured losses of this kind, especially at the field level. One reason for this absence of information is the problem of experimental control; rarely does a pest or a disease act independently in causing yam crop losses. Attack by one pest or disease is an invitation for others to join; under such a situation it is hard to ascribe the yam crop loss to a single pest or disease, whether in the field or in storage. Damage to yam in particular and to the ecology in general caused by the problem of nematodes in yam should be defined broadly to include the contributions of the nematode problem to the high cost of seed yam and to low adoption of various yam technologies by farmers.

The high cost of nematodes in the yam food sector includes the continued practice of yam production under extensive methods with negative ecological impacts. Extensive methods are defined for the present purpose in terms of slashing and burning, opening rotation cycle with a crop, fallowing period, and number of times a crop is grown in the same plot before changing to another crop or to fallow. Slash and burn is a thermal treatment that helps to clean the soil, debris, and alternate plant hosts of yam pests and diseases. Alternating growing yam with fallow or with another crop on a plot helps break the pest and disease cycles, while virgin land is relatively slow in the buildup of yam pest and disease problems. Therefore, extensive production practices are used as control measures against yam pests and diseases, but such practices are costly in terms of inefficient use of land, labor, and other resources as well as being environmentally unfriendly (Thrupp et al. 2006).

The high costs of nematodes in the yam food sector also include the narrowing of yam production areas to limited ecological zones. The West African yam belt is defined in the south, east, and west by high humidity that favors proliferation of some of the most damaging yam pests and diseases, such as the nematode pests. Facing up to the challenge of nematodes in the yam food crop sector will expand the yam belt in the southern, eastern, and western areas of West Africa where these problems hold back the production of the crop.

The problem of nematodes also limits rural assembly markets to a minimum in the marketing channel, drastically reduces speculation marketing functions, and discourages bulk buying of yam for home consumption because traders and consumers are intimidated by the nematode problem. In addition, the yam nematode problem is incriminated in sidelining yam in national food policy programs in various countries across West Africa. Because of low adoption rates of various yam technologies by farmers, government farm input subsidies that are based on new technologies, including a subsidy on quality-declared seeds, are available for grains but are not extended to yam.

Pain Killer Solutions for the Nematode Problem in Yam

Researcher-recommended control measures for nematodes in yam are thermo-
and chemotherapies. These treatments are completed by washing the seed yams
in hot water or in the appropriate chemical solutions (Nwauzor 1998b). The
research-recommended measures do not constitute effective solutions; they are
"pain killer" treatments that produce the washed pig effect under field conditions.
The realities of the field conditions include soil and debris, which are dense with
the nematodes, a wide range of uncultivated weeds, and cultivated intercrops,
which play host to the nematodes. Nematodes easily spread from fields planted
with nematode-infested seed yams to fields planted with nematode-free seed yams.
Comparable to a washed pig, which becomes soiled once back in her sty, thermo- or
chemo-treated seed yam is reinfested in the field and after harvest and carries the
same problems to the store.

Challenges to Finding an Effective Solution

What will constitute an effective solution to the problem of the nematodes of yam?
At the Progress Monitoring and Discussions Workshop of IITA's YIIFSWA project
held in Kumasi, Ghana, from April 28 to May 4, 2013, Dr. Lava Kumar, IITA plant
virologist, quoted Richard Gibson as saying that resistance breeding holds the key
to effective control of crop pests and diseases.[2] The challenge of yam breeding for
resistance against nematodes is profound. Nematode resistance genes have not
been identified in cultivated yams. Recourse to wild relatives of yam as sources of
such genes is being pursued with caution by breeders because some of these wild
relatives contain lethal substances.

In addition, crop breeding, especially breeding of a complex crop such as yam,
is a time-consuming effort, and breeders are, in most cases, employed on short- or
uncertain-term contracts. At best, yam breeders are enticed out of breeding by
promotions to top administrative positions in their institutions. In 2006 Dr. Robert
Asiedu, who effectively competes with Dr. S. K. Hahn as the leading yam breeder
worldwide, was promoted to the high position of research for development director,
West Africa at IITA. Because of his passion for yam breeding, Dr. Asiedu continued
to breed yam for five years until 2011 when he was advised by IITA authorities to

focus on his administrative job. That year Dr. Antonio Lopez-Montes, a Colombian national, was appointed as yam breeder/geneticist at the IITA.

In 1994, Dr. Margret Quin, director of the IITA's Crop Improvement Program, encouraged Dr. Jacob Mignouna, a biotechnologist/plant geneticist at the IITA, to focus his research effort on providing support for yam research. Dr. Mignouna initiated systematic work to understand the genetic structure of cultivated and wild species of yam in West Africa. Dr. Mignouna collected yam germ plasm from all over West Africa, characterized their genetic diversity, and developed the first genetic mapping populations for white and water yams. He then determined the genetics of the mosaic virus disease in white yam and anthracnose in water yam and mapped the genes that confer resistance to those two diseases (Mignouna et al. 2002). In the year 2000, when he was close to achieving a breakthrough, Dr. Mignouna's contract at IITA was abruptly terminated.

Call for Radical Solutions

Formidable impediments to resistance breeding suggest that resorting to certain radical measures, such as biological control and transgenic engineering, to find a solution to the nematode problem in yam may be necessary. Effort has been initiated in the context of student work in the case of biological control solutions. But the experience of the successful cassava mealybug control program in Africa demonstrates that a biological control program includes the critical roles of global partnerships and cooperation of multiple participants and countries in tackling complex problems (Neuenschwander 2003). For example, in order to tackle the cassava mealybug problem, an Africa-wide biological control center was established at the IITA station in Benin Republic. The IITA brought together a group of international scientists and donors who crisscrossed Central and South America and eventually found a parasitic wasp that kills mealybugs in the process of reproduction without causing damage to any crop, animal, or ecology at large.

Transgenic engineering is another radical measure that ought to be intensified, especially because resistance genes are proving difficult to find in cultivated yams and some wild relatives of yams contain poisonous substances. Application of the tool on cassava is in progress under the leadership of the Donald Danforth Plant Science Center that is addressing the problem of low contents of micronutrients in cassava.[3] People who are outraged by the prospect of genetically modified organisms (GMOs) in Africa are privileged enough to have full stomachs. Hunger is the first line in the arithmetic of poverty because hungry people are unable to work to improve their income. Better than international standards of comparison, such as a dollar per person per day, the depth of poverty in the region is appreciated through dismal stories returned by field-level R&D personnel who, by the nature of their work, possess a front-row view of poverty.

Dr. Regina Kapinga reported that in a community in Tanzania, some parents would not pay for a school feeding meal of maize porridge at the cost of US$0.70 per child per *month* because those parents had more pressing needs for that amount of money.[4] Dr. Djana Mignouna reported a story he heard told by HIV/AIDS workers among promiscuous herdsmen who had a high rate of infection.[5] The workers asked if the herdsmen used condoms, and they responded that, yes they used them. The workers asked to see one and the herdsmen produced something like an enlarged small latex sack with a hole at the tip, the only one still left in the village for use by anyone. One could use the condom on the condition that after use it is washed and dried in the sun as a courtesy to the next user. The incredible school lunch story is real; the condom story may have been inspired by the reality of poverty on the ground in Africa. People who struggle with poverty so deep, people who are unsure of what their little children will eat in the next few days, do not raise an outcry against GMOs. But voices of opposition to GMOs should be listened to in order to take necessary precautionary measures.

Summary

Nematodes are one of the biggest obstacles in the development of the yam food sector. Besides causing yield losses, the problem

1. holds back the diffusion of various yam technologies including novel seed yam technologies,
2. restricts yam production expansion in marginal humid ecologies,
3. prevents yam cultivation by intensive methods,
4. reduces yam marketing efficiency, leaving producers helpless at the price negotiation table with the yam traders,

5. adversely affects yam consumer decisions, and

6. renders national food policy programs irrelevant for yam.

Present research-recommended control measures are palliative and therefore not effective. It has been suggested that resistance breeding holds the key to effective control of various crop pests and diseases, but nematode resistance genes have proved difficult to identify in cultivated yams, while some wild relatives of the cultivated yams that may provide resistant genes contain lethal substances.

The situation calls for urgent attention! The experience of the successful cassava mosaic resistance breeding in Africa demonstrates that nematode resistance breeding is a time-consuming endeavor. Young scientists need to be encouraged to commit their life's work to finding a solution to the problem by receiving necessary, long-term job contracts. Older scientists may be discouraged by the time it takes to achieve breakthrough. The cases of Drs. Asiedu and Mignouna illustrate the helplessness of scientists working to make an impact when administrators have a reason to halt their promising work by terminating their contracts or by luring them away from yam breeding with promotions to top administrative positions.

The absence of nematode resistance genes in cultivated yams and the presence of lethal substances in wild relatives of yam are slowing the progress in finding a solution through resistance breeding. This suggests that radical solutions may have to be resorted to. A biological control program is one possibility; the program is radical because it is complex and, as the lesson of the successful program for biological control of the cassava mealybug in Africa teaches, it involves numerous institutions and countries. Transgenic engineering is another radical solution. Voices of protest against it are useful; they will help keep scientists focused on important security issues.

CHAPTER 6

Yam Crop Improvement Research

I n West Africa, yam crop improvement research is conducted as a mutually beneficial effort between IITA and various national research centers. In the national research centers the now well-known logistics problems of research centers in Africa such as inadequate research infrastructure and supplies apply to yam research programs as well. Those problems are overcome by collaboration with IITA, which allows the researchers of the national programs access to global scientific information, equipped laboratories, and a germ plasm collection, as well as the critical mass of scientists in varying disciplines required for effective crop improvement research. On the other hand, in absence of the collaboration, IITA could conduct necessary germ plasm evaluation trials across a wide range of ecologies at high costs.[1]

Establishment of Yam Research Programs in Anglophone West Africa

In 1893, the British colonial government established a botanical research station in Lagos, Nigeria, and there the colonial botanists achieved a measure of success in characterizing the morphology of the yam vine and tuber, as well as the plant's flowering habits (Akoroda 1992; Olayide 1981). In 1956, a formal research program on yam was set up in Nigeria following a national economic survey in 1951 and 1952 that revealed the high value of yam in the economy (Waitt 1961). In Nigeria, early efforts in yam crop improvement research were making collections of cultivated and wild varieties from South Africa, Ghana, Central America, and from within the country; the acquisitions were maintained in Umudike near Umuahia in eastern Nigeria (Waitt 1961).

In 1971, a yam research program was started at IITA as part of the institute's Root, Tuber and Plantain/Banana (*Musa*) Improvement program led by Dr. S. K. Hahn, a Korean breeder who joined the institute in the same year. In 1972, Hahn helped establish a formal yam research program in Ghana under the leadership of Dr. John Otoo. Before the establishment of these yam research programs, yam research was being conducted informally in the various universities of Nigeria and Ghana.

Absence of Genetic Foundation for Yam Breeding

By the time formal yam research programs were established in Nigeria in 1956, in 1971 at IITA, and in 1972 in Ghana, it was already known, through earlier work of botanists and university researchers, that the poor flowering habits of yam would constitute a profound impediment to yam crop improvement research. The botanists had established that, although some varieties and species of yams were known to flower and to produce viable seeds in nature, yams were unpredictable in terms of regularity and intensity of flowering and in terms of seed setting. Some varieties do not flower at all, while others may flower in some years and fail to flower in other years. Male and female yam plants differ in flowering intensity and in their periods of flower initiation. Male yam plants flower more intensely than females, and there is poor synchronization of flowering periods: male plants flower early, while female plants flower later (Segnou et al. 1992; IITA 1983). This was the only genetic information that the newly established yam research programs had available for their research.

Prospecting for Germ Plasm and Taming Cultivated Yam Varieties in Terms of Flowering Habits

The germ plasm collected and maintained in eastern Nigeria by the Nigerian yam research program, which was established in 1956, was lost during the Nigerian Civil (Biafra) War (1967–70). With the knowledge that the poor flowering habit would pose a formidable challenge in yam crop improvement research, yam researchers embarked on a germ plasm collection expedition with the initial aim of improving the flowering habits in cultivated yams. Most of the world's yam germ plasm was found in West Africa because that was the center of origin of the cultivated white yam species. The researchers collected germ plasm of cultivated and wild yams from both within and outside of West Africa. Caution was exercised in the use of wild relatives as breeding materials because some of them contained lethal substances.[2]

The next tasks of the yam breeders were to induce flowering in nonflowering cultivated yam varieties and regularize and synchronize flowering in male and female plants. With persistent prospecting for local landraces, researchers at IITA and in the two countries, that is, Ghana and Nigeria, established an elite collection of yam germ plasm that had good flowering tendencies. Working from this genetic base, the researchers were able to shed light on the many interlinked crop management and environmental determinants of flowering and seed-setting behavior in commonly cultivated yams. Among these factors are the time of planting, the type of planting material (seed or tuber setts), the size of the sett, soil moisture, atmospheric humidity, soil fertility, and photoperiod. With much field experience and a better understanding of biological and environmental processes at work, researchers were set to manipulate yams into flowering and seeding.

In 1974, the technical feasibility of artificial hybridization was established when researchers artificially crossed local female and male white yam. That was a major success for yam research as it was the first time cultivated yam was artificially crossed, although the value of hybridization is found in the crossing of different cultivars to take advantage of their different genetic attributes (IITA 1996). Independent reports of germination of true seeds of the white yam sett in nature by Okoli and by Sadik and Okereke swelled the success of artificially crossing female and male white yam (Okoli 1975; Sadik and Okereke 1975).

Hybridization and Clonal Selection

The early solution to the puzzle of yam flowering habits, just three years after the establishment of the IITA's yam research program in 1971, paved the way for yam breeding in West Africa. One breeding goal was to develop cultivars that have attributes including high tuber yield, pest and disease resistance, storability, and culinary quality. Another breeding goal was to develop cultivars with characteristics that lower the labor requirements, eliminate staking, and reduce the amount of planting material required to produce the yam crop (Okoli 1980). These objectives were derived from the problems of production, namely high seed cost, high labor cost, and pest and disease problems (Ene and Okoli 1985).

To achieve these objectives yam breeders at IITA led by Dr. S. K. Hahn embarked on ambitious breeding programs. With the flowering puzzle solved, poor yam seed sett and low seed viability were the next bottlenecks of breeding to fix. The low rate at which seed yam can be multiplied for clonal evaluation and selection is another difficulty faced in yam crop improvement research (Waitt 1961). These outstanding bottlenecks required that a large number of crosses be made to generate sufficient viable seeds for clonal selection. Literally thousands of crosses were made each year by Dr. S. K. Hahn and his team of researchers to try to achieve these goals. In 1989 IITA began distributing yam clones they had bred to several national agricultural research centers within and beyond West Africa for clonal selection (IITA 1995).

From 1988 to 1991, Dr. Larry Stifiel, director general at IITA from 1986 to 1991, stopped the yam breeding research program by the IITA. At this point there was high optimism among yam researchers because of their recent breakthroughs in taming commonly cultivated yam varieties in terms of flowering habits, germinating true seed of yam sett in nature, and artificially pollinating white yam varieties.

Dr. Stifiel started setting priorities in IITA's program in terms of geographical and commodity coverage to rationalize cost and produce impact within a reasonable period of time. In 1986, Dr. Stifiel hired Dr. Paul Dorosh, an American postdoctoral fellow, to assess root, tuber, and *musa* crops for relative importance and potential return on investment in research. Dr. Dorosh based his assessment on the FAO's agricultural production and consumption data and prepared the report "The Economics of Root and Tuber Crops in Africa" (Dorosh 1988).

The report highlighted the high cost of yam relative to cassava and yam's positive income elasticity of demand, noting that as per capita incomes rose, people tended to consume more yam. The report contributed to the decision to "Phase out breeding research on yams" (IITA 1989, 157). This decision was made without understanding that unlike cassava, which had been researched since 1891 beginning at the East Africa Agricultural and Forestry Research station in Amani, Tanzania, formal yam research hadn't started in Africa until eighty years later in 1971. The decision failed to consider that yam consumption by a high-income population segment points to its potential for high returns on investment in research. It also failed to consider that improved productivity through research would raise producer incomes through reduced production costs and consumer incomes through reduced yam prices to the advantage of the low-income population segment.

With the departure of Dr. Stifiel in 1991 the yam research program was restored at IITA. Dr. Robert Asiedu, a Ghanaian researcher, was redeployed from cassava to work as breeder in the resumed yam program. Dr. Asiedu pursued the aggressive yam improvement work of Dr. Hahn with enthusiasm that mirrored vengeance for the time lost from 1988 to 1991. A story often told in yam research circles in Africa is that Dr. Asiedu planted more yams than he could harvest. In 1992 Dr. Asiedu made many crosses and generated a considerable amount of true seed. He raised them in a nursery and planted them out in the field for an evaluation trial at the IITA headquarter station in Ibadan. At the end of the season the entire planting could not be harvested; it was too much.

In 2001, after thirty years of painstaking work in germ plasm collection, flower induction and synchronization, crossing, and clonal selection trials, three hybrid yam varieties were officially released to farmers in Nigeria. Since then sixteen more varieties have been released in Nigeria (appendix 2). The released "hybrid yams performed significantly better than the landraces with the exception of DRN010 in terms of uniform and early sprouting, crop establishment, vigor, survival up to harvest, total tuber yield, seed yam and ware yam productions" (Orkwor et al. 2001, 353).

Outstanding Challenges in Yam Crop Improvement Research

The superior performance of the released hybrid yam varieties in terms of tuber yield can be considered genetic potential for high yield. The attainment of this high yield depends on a wide range of farm-level factors, especially high pest and disease incidence, which depresses yield. Set against the backdrop of formerly stated yam breeding goals, the released varieties fall short in some of the hoped for attributes, including pest and disease resistance, characteristics that lower labor requirement, and characteristics that reduce the amount of planting material required for producing yam crop.

Summary

Formal yam crop improvement research in Anglophone West Africa started in the early 1970s when the IITA's yam research program was established in 1971 and Ghana's was established in 1972. The Nigerian program, established in 1956, collected germ plasm that was later lost in the Nigerian Civil (Biafra) War. The newly established yam research programs at IITA, Nigeria, and Ghana had only morphology and tuber characterization information provided by botanists as the genetic resource to start with for their yam crop improvement research. The information was an important resource because it prepared the researchers in the newly established programs for the profound challenge posed by the poor flowering habits of yam to their breeding work.

Within three years of establishment of the programs, the researchers achieved the breakthrough of inducing flowering in major cultivated yam varieties and synchronizing the flowering period of male and female yam plants, setting the stage for an ambitious yam breeding program. From 1988 to 1991, IITA halted the yam crop improvement research at a point of high optimism raised among yam researchers by their recent breakthroughs. The breakthroughs included taming commonly cultivated yam varieties in terms of their flowering habits, germinating true seed of yam sett in nature, and artificially pollinating white yam varieties. In 1991 yam crop improvement research was restored and researchers continued pursuing crossing and evaluation with a considerable level of aggressiveness. But still, it took thirty years from the establishment of the research programs in the early 1970s to make the initial releases of new yam varieties to farmers in 2001.

What insights can be gained from these analyses? Successes achieved in the yam crop improvement research, some within a short time period, are convincing evidence that yam responds positively to investment in research. But it took a long time—three decades—to come up with new varieties, especially because there was no genetic material developed earlier to build on. Pest and disease problems constitute the most critical challenge in yam crop improvement research. Young scientists should be encouraged to build their career on finding solutions to it as the problem may not be attractive to older scientists. The older scientists may be uncertain that a solution would be found within their professional life cycle, deterring them from taking on the challenge. Researchers often do not have control of their work schedule as administrators can, at any time, stop the researchers' work. In addition, successful researchers are often enticed out of scientific research by promotion to high administrative positions.

CHAPTER 7

Prospects and Impediments to Diffusion of Hybrid Yams in Nigeria

What is the level of adoption of the hybrid yams developed jointly by the NRCRI and the IITA after more than ten years since the initial release to Nigerian farmers in 2001? During the yam consumption study, farmer groups interviewed in the leading yam-producing centers in the country were asked to rank, in descending order, the five most common yam varieties grown (see table 7), as well as how much was grown of each, about what year each was first grown in their area, and from where it was introduced. Each farmer group included the oldest yam farmer in the village. The answers to this set of questions suggest that the approximate dates of introduction of most commonly grown yam varieties in the areas are not known to living farmers.

Two varieties were reported to have been introduced in the last ten years before 2013, both of which came from other towns where they were known to have been grown for many years. This means that most of the hybrid yam varieties, if any, have not made the list of the five most popular varieties. The possibility still exists that farmers are unknowingly growing some of these varieties because yam varieties express different morphological characteristics under different ecological conditions (Waitt 1961). Also, through ethnobotanical, morphological, and molecular analyses Emmanuel Otoo and others show that one particular variety of the *pona*,

TABLE 7. Nigeria: The most commonly grown yam cultivars in major yam centers (in order of popularity), December 2012

YAM ZONE	CULTIVAR	YAM ZONE	CULTIVAR
Shaki	Kokoro (Ihobia)		Agbocha
	Amula		Abii
	Lasiri	Zaki Ibiam	Ogoja
	Kemi		Danacha
Otuocha	Ekpe		Gbango
	Adaka		Sampeper
	Obiaoturugo		Gisa

Source: Data from Nweke, Aidoo, and Okoye, 2013.

a popular white yam variety in Ghana, is genetically a composite of some other white yam varieties (Otoo et al. 2008).

Pest and Disease Susceptibility

Descriptors of ten of the nineteen varieties of hybrid yams released in Nigeria between 2001 and 2010 listed "pest and disease tolerance" among outstanding characteristics. However, this attribute was absent for the remaining nine varieties (appendix 2). This generalized information, which suggests that the ten hybrid yams are tolerant to all yam pests and diseases, is an insufficient basis for judging the superiority of the hybrid yams over local varieties in terms of the important attribute of resistance to pests and diseases.

Dr. G. C. Orkwor and others provided more details for the first three varieties that were released to farmers in 2001: "In terms of resistance to pests and diseases, the hybrid yams appeared more susceptible to yam nematode attack, anthracnose and beetles than the land races especially at Umudike in 1997. The hybrid yams however were less susceptible to virus attack than the landraces" (Orkwor et al. 2001, 353).[1] Why should researchers release to farmers yam varieties that at the point of release appear to be more susceptible to pests and diseases of economic importance such as yam nematodes, anthracnose, and beetles? Dr. Robert Asiedu, a yam breeder at IITA for many years, explained that a new variety is released to farmers if it has attributes that are superior to existing varieties and not inferior in any way.[2] This explanation is an insufficient justification for the release to farmers of the hybrid yams because they appear to be inferior to local varieties in terms of resistance to some pests and diseases of economic importance.

The wisdom of releasing to farmers a hybrid yam that is susceptible to any of the yam pests or diseases is questionable, unless an alternative solution to the pest and disease problems is already in place. This position is informed by an experience reported by a Kintamkpo group of farmers interviewed in central Ghana during the yam consumption study. The group of farmers reported that a few years earlier a white yam variety was introduced to them from another yam growing area. It performed better than any of their existing varieties in terms of tuber size and shape and culinary quality, but within a few years of growing, the introduced variety collapsed because of a pest or disease and its cultivation was abandoned.[3]

The diffusion experience of the IITA's initial high-yielding, mosaic-resistant TMS cassava varieties teaches a lesson about diffusion of new crop varieties. Delivering new varieties of vegetatively propagated crops to farmers requires large investments in seed multiplication and distribution. This is because the vegetatively propagated crops have low multiplication rates and are perishable with a short shelf life, and they are bulky and therefore expensive to transport. Seed yam, the vegetative propagation material, is the edible tuber, that is, the crop; it is more expensive in terms of price per unit weight than table yam and even ceremonial yam tuber. Seed yams are easily exchanged for cash; free distribution has a high likelihood of abuse because everybody, farmers and nonfarmers, would like to receive free seed yam for planting, sale, or consumption.

The decision to invest in multiplication and distribution of the hybrid yams at the present stage of the yam crop improvement research should be made with caution. The susceptibility of the hybrid yams to pests and diseases could lead to an embarrassing failure that can create situations of reverse adoption, farmer disaffection with research, and loss of funds invested unless an alternative solution for the problem is found.

Ecological Adaptation

As in the case of pest and disease susceptibility, the descriptors of the hybrid yams are light with information on ecological adaptability. For example, the descriptors state "stable yield" as an outstanding characteristic for the seven hybrid yams released to farmers in 2001 and 2003, without elaboration regarding ecological zone or time duration over which the yield is stable (appendix 2). For the remaining twelve hybrid yams released after 2003, the descriptors are silent on the matter of ecological adaptation.

Again Dr. G. C. Orkwor and others provided more detail on the first three hybrids released to farmers in 2001; Orkwor and his colleagues state, "forest, S. G. [southern Guinea] savanna and savanna" under adaptation (Orkwor et al. 2001, 355). This is a sweeping generalization on more than one ground; the generalization is based on "at least seven locations for two years coupled with the results of the on-farm testing and farmer's reaction and evaluation of the hybrid yams" (Orkwor et al. 2001, 355).

Nigeria is a country with a large land mass, a large proportion of which is part of the yam belt of West Africa. Information collected from test experiments in seven locations is insufficient to conclude that the hybrid yams are adaptable to "forest, S. G. savanna and savanna," which cover the major yam-growing ecologies of the country. That the three hybrid yams released to farmers in 2001 can be equally adapted to all major yam-growing ecologies is too much good news. Two years of test experimentation for a crop that is grown only once in a year—because of its long growth cycle—is also inadequate for the conclusion that the hybrid yams are adaptable to "forest, S. G. savanna and savanna."

The implication of this generalization is that the information available is insufficient for implementing measures to promote the diffusion of the hybrid yams. Clear knowledge of which hybrid yam is suited for promotion in each agroecology is precondition for successful implementation of a diffusion program.

Lack of Investment in Adoption Promotion

This section draws on the history of some agricultural technologies' diffusion. These agricultural technologies were chosen because they achieved varying levels of diffusion in Africa. This will show that, while investment in promotion is necessary to speed up diffusion, good technologies are able to spread on their own, even if slowly, provided that preconditions for their adoption, if any, are in place. Most technologies require precondition for adoption; availability of chemical fertilizers and irrigation water were preconditions for adoption of the Green Revolution wheat, rice, and maize in South America and Asia in the 1960s and

1970s. A mechanical cassava grater for preparing *gari* was in place in Nigeria when the IITA's high-yielding, mosaic-resistant, TMS cassava varieties were developed and released to farmers in the late 1970s (Johnson and Masters 2004). Without the mechanical grater, farmers would be hard pressed to secure the sufficient labor needed to process, by hand, the amount generated by planting the high-yielding varieties. Control of major yam pests and diseases such as nematodes, anthracnose, etc. is a precondition for diffusion of high-yielding hybrid yams in Nigeria.

The COSCA studies in Nigeria and Ghana revealed that the IITA's high-yielding, mosaic-resistant, TMS cassava varieties were being grown in some cassava-producing villages before the varieties were officially released to Nigerian farmers in 1977 and Ghanaian farmers in 1993 (Nweke, Spencer, and Lynam 2002). The high-yielding, mosaic-resistant, TMS cassava varieties scaled the research institution fence through research assistants and began to spread without promotion because the varieties were resistant to cassava mosaic disease, which was a major problem of cassava production in the two countries. It also began to spread without promotion because the laborsaving mechanical cassava grater for preparing *gari* was already in place. Promotion by multiplication and free distribution of planting setts by the government later surged the momentum of the already ongoing diffusion process.

Hybrid yam varieties must have also scaled the research institution fence through the research assistants and especially through farmers during the on-farm adaptive trials on the farmers' fields. Unfortunately, they either were eaten or, if planted, did not spread because the pressing problem of yam pests and diseases was not solved. They probably collapsed in the fields of farmers under the pressures of yam nematode, anthracnose, and beetles to which problems they were more susceptible than local varieties.

An introduced water yam variety, namely the Florido yam (*Dioscorea alata*), variously called *asana*, *matches*, and *sudan*, in West Africa is an example of a technology that initially spread on its own. This was because it solved farmers' felt needs and did not require a precondition for adoption. The diffusion was later sped up by promotion. Sékou Doumbia, Muamba Tshiunza, Eric Tollens, and Johan Stessens investigated the factors that influenced the rapid spread of the variety in Côte d'Ivoire where it was introduced from Puerto Rico in the early 1970s by the Institut de Recherches Agronomiques Tropicales for a yam production mechanization project (Doumbia et al. 2004). Doumbia, Tshiunza, Tollens, and Stessens stated that for various reasons the mechanization project was abandoned, but the variety was nevertheless taken up by farmers in the late 1970s. It was rapidly adopted and, since then, has spread throughout the country.

The major reasons for the success of Florido among Ivorian farmers include its flexibility in the rotation cycle and planting time, the good storability and extended storage life of its tubers, and its resistance to most pests and diseases. This is especially true for the internal brown spot disease and wilting, which usually adversely affect the production of water yam varieties (Doumbia et al. 2004). Doumbia, Tshiunza, Tollens, and Stessens concluded that "It is clear that a variety introduced into a country for the wrong reasons can nevertheless spread rapidly if it has superior technical and economic attributes. Moreover, take-up of the crop may not require the active support of the country's national agricultural research and extension system" (Doumbia et al. 2004, 49). The lesson is that the hybrid yams do not need to wait to be promoted for more than ten years after the initial official release to farmers in order to spread if they are a solution to the pressing producer problem of yam pest and disease susceptibility or if an alternative solution to the problem is in place.

Later, the Florido water yam variety was aggressively promoted in Nigeria, Benin, and Togo by different R&D agencies.[4] The Sudan United Mission in Abakaliki multiplied and distributed the variety in eastern Nigeria under the leadership of Hans Meerman. The IITA, in collaboration with NRCRI, multiplied and distributed the variety in the south geopolitical zone of Nigeria through the Green River Projects, and Dr. Philip Vernier distributed truckloads of the variety in Benin and Togo under the yam valorization project.[5]

Pona, a white yam variety, is popular with consumers in Ghana and other West African countries where it is produced; it is also the most widely traded yam in the international yam market. It has a drawback, namely a short shelf life; it is more susceptible to postharvest virus and fungus attacks than most other local varieties. In spite of this, consumers patronize it with premium price because it is superior to most other local white yam varieties in terms of palatability, texture, and fast cooking attributes. The *pona* case is comparable to fruits and vegetables like papaya and spinach, which consumers patronize for other reasons than keeping quality. Knowing that those commodities are highly perishable, consumers use them shortly after purchase.

The history of *pona*'s introduction to its growing areas is unknown; there is no written record, and the yam consumption study revealed that nobody alive was able to tell us when or from where it was introduced into the various growing areas.

This means that farmers have continued to produce it over many years because it matures early and its yield has been sustained at a profitable level over its long history of cultivation. In comparison, it is uncertain at this stage whether the hybrid yams will sustain yield at a profitable level since they are more susceptible to some major yam field pests and diseases than local varieties and as their yield stability over time and across agroecologies is uncertain.

The minisett seed yam multiplication technique is a good technology because it has the potential to solve the felt problem of a slow multiplication rate. The technique has been widely promoted through extensive on-farm trials in several agroecologies and under various government programs and projects including the National Accelerated Food Production Program and the Root and Tuber Expansion Program in Nigeria and the Root and Tuber Improvement Program and the Root and Tuber Improvement and Marketing Program in Ghana. The technique has been investigated over and over with recommendations for modifications to make the technology adoptable.[6] But evidence found in literature on the subject suggests that after some forty years since development in the early 1970s, the technology has achieved a low adoption rate; it was found to be unprofitable.[7] The reason for the unprofitability and the consequent low rate of adoption is that the precondition for its adoption—control of yam nematodes, viruses, fungi, etc.—is not in place.

Prospect for Diffusion of Hybrid Yams in Nigeria

The lesson of this chapter is that the ability of a new technology to solve a felt need and the preexistence of complementary technologies (preconditions for adoption), if any, are necessary conditions for adoption; promotion serves to speed up momentum of diffusion. More than promotion, the prospect for diffusion of hybrid yams in Nigeria depends on the establishment of precondition for adoption, namely control of major yam pests and diseases, especially nematodes, which constitute the biggest impediment against adoption of the hybrid yams.

CHAPTER 8

Periphery Situation of Yam in National Food Policy Programs

T he situation of yam at the periphery of national food policy programs in West Africa is illustrated with Ghana. The situation has been evident in Ghana's food policy from the early 1960s, after independence, to the present time.

Direct Food Production by State Agencies

The Republic of Ghana became politically independent of British rule in 1957. From 1962 to 1966, the government of Ghana adopted a socialist policy, which recognized that "the need for the most rapid growth of the public and cooperative sectors in productive enterprises (agriculture and industry) must be kept in the forefront of government policy" (Ghana 1964, 2). In those same years, several public agencies were established for the purpose of achieving this stated policy goal. For example, the State Farms Corporation (SFC) was established to do those things that, in the opinion of government experts, the private farmers could not be relied upon to do—to use modern methods to expand the production of food crops and raw materials on a commercial scale (Ghana 1971). The SFC took over the agricultural experimental stations from the Ministry of Agriculture, converted

TABLE 8. Ghana: Area and yield of staple crops in state farms compared with national levels, 1974

CROP	STATE FARMS AREA (HA)	NATIONAL TOTAL AREA (HA)	STATE FARMS AVERAGE YIELD (TONS/HA)	NATIONAL AVERAGE YIELD (TONS/HA)
Cassava	1,283	378,947	13.59	15.68
Millet	338	221,862	0.76	0.75
Sorghum	405	214,980	0.79	0.95
Maize	3,941	413,765	1.24	1.35
Rice	1,350	66,000	1.58	1.23
Total	7,317	1,295,554	—	—

Sources: Data for state farms are from Food Production Corporation, Annual Report 1974. National data on area and yield for cassava, millet, sorghum, and maize are from Ministry of Agriculture (ERPS) Report on Current Agricultural Statistics, Accra 1994; and for rice from U.S. Department of Agriculture.

them into production units, and established additional farms. Other agencies established by the government to engage in food production include the United Ghana Farmers' Cooperative Council and the agricultural wing of the Workers' Brigade (Ghana 1971).

By 1966 the state production agencies were operating a total area of about 7,000 ha. This included virtually all major food staples except yam, plantains, and cocoyam, while the yields achieved were not different from smallholder levels (table 8). Why did the state agencies exclude yam in their production? Yam is not amenable to production on a large scale because seed yams are not available in large quantities even in open markets and soil conditions suitable for growing yams are available in small niches. Additionally, available tractor mechanization technologies are not suitable for performing various yam agronomic tasks such as mound making, staking, weeding, and harvesting.

The first independent government was taken over in a coup d'état in February 1966. The policy in agriculture in 1967–75 was redefined to encourage increased participation by the private sector in such areas as the distribution of farm inputs; the Ministry of Agriculture would continue and in 1970–71 largely complete the process of disengagement from the earlier policy of state-controlled production (Ghana 1971). Since then Ghana has gone through political, economic, and weather-related crises that have affected agricultural policy programs. The latest of the crises was the severe drought of the early to mid-1980s.

Farm Input Subsidy Programs

Food and Agriculture Development Policy II listed maize, cassava, rice, yam, and cowpea—in that order—as priority crops for policy support in Ghana (Ghana 2009, 2010). Farm inputs made available to farmers at subsidized prices were quality-declared seeds, chemical fertilizers, farm machinery, irrigated farm land, and farm credit. The aim was to increase the production of the priority crops.

How much of each of the subsidized inputs is made available to farmers?[1] The quantity of chemical fertilizers available in the country was increasing; it increased from about 4 kg/ha of arable crop land harvested in 2002 to about 20 kg/ha in 2009 (FAOSTAT).[2] This is too low to have an impact on yield on a national scale. If the entire amount of fertilizer available in 2009 was used for the priority crops alone it would amount to about 35 kg/ha, which is well below the research-recommended levels of 200 kg of NPK 15:15:15 per ha for rice or 400 kg of NPK:15:15:15 plus 150 kg of urea as top dressing per ha for maize, for example.[3]

Subsidized farm machineries include tractor mechanization and mechanical maize shelling machines. From 1976 to 2005, the number of tractors available in the country was around two thousand, while the area of priority crops harvested nearly tripled from about 800,000 ha in 1976 to about 2,200,000 ha in 2005 (FAOSTAT). Clearly the number of tractors available during the period was too low to make an impact at national level in terms of reduction in labor input in the production of the priority crops.

Information on the number of maize shelling machines was not available, but those machines were numerous and visible in periurban centers, especially in major maize-producing regions such as the Northern, Upper Eastern and Upper Western Regions, and parts of the Brong-Ahafo region. The shelling machines were initially introduced and subsidized by the government, but they became popular with the maize farmers. Now they are fabricated and maintained by local artisans; they are mobile either by mounting them on small trucks or by pushing them on their own wheels; they are run by petroleum fuel and are therefore used in rural areas without electricity supply. One machine is sufficient to serve the needs of a collection of contiguous villages, even at the peak of harvesting season. Entrepreneurs own the machines and provide shelling services to smaller farmers.

Although the government may have committed substantial amounts of funds in irrigation development since the early 1960s, subsidy is ineffective because the amount of land area equipped for irrigation is insignificant in comparison with the national land area of any one of the priority crops. Land area equipped for irrigation was 6,000 ha until 1996 when it rose to 34,000 ha in 2007 and remained at that level by 2009 when data was available. At this point, the area of maize harvested was approximately 790,000 ha; rice, 108,000; cassava, 800,000; and yam, 348,000 (FAOSTAT).

The subsidized inputs are distributed to farmers through two channels. One channel is the open market where any farmer—including producers of nonpriority crops—who is willing and able to pay can buy inputs at subsidized prices where and when the inputs are available. The second channel is through farmer-based organizations, namely block and cooperative farmers. School leavers are brought together to form block farmers, and individual farmers are encouraged to form cooperatives in order to benefit from the various subsidy programs.[4] Large tracts of farmland are cleared, plowed, and blocked out to the school leavers and to members of cooperative societies in portions of a few hectares to each farmer for the purpose of producing mostly rice and maize. The inputs are provided with a zero interest loan that is repaid with crops at harvest, which is used as stock for food shortage buffering.

Obviously, yam does not benefit from these input subsidies, whether distributed through open market sales or through block farms except in a few cases where staking conditions permit yam field plowing with tractor mechanization. Farmers rarely irrigate or apply chemical fertilizers to yam because of their conviction that irrigation and fertilizers expose the crop to nematodes, viruses, and fungi. Available seed yam technologies have proved inefficient for large-scale multiplication.

Road Infrastructure Development Program

The degree of road infrastructure development at farm level is difficult to assess. Yam producers are faced with the challenge of time-consuming and burdensome daily commute over long distances between settlements and yam fields in the yam food sector. The distances increase each year as the farmers move further and further into distant forestlands in search of sites with fertile soil, stake trees, and low yam pest and disease pressure.

The continuous movement by thousands of smallholder farmers in different directions to distant forestlands year after year handicaps the use of the road infrastructure development program to promote yam production. New yam production technologies, which will permit yam production by intensive methods and discourage farmers' constant movement into distant forest areas in search of virgin lands, will facilitate implementation of the government's road infrastructure development program to promote yam along with other priority food crops.

Food Shortage Buffering Strategies

Among imported priority crops are rice and, to a lesser extent, maize; other priority crops, especially yam and cassava, are not available for importation in the international market. Between 1961 and 2009, maize importation per capita was insignificant in comparison with rice importation (figure 5). Stockpiling was low because the farm credit that was linked to the input subsidy on which it depended was inefficient (figure 6). The inputs supplied to the farmers on credit were small compared to national need. The rate of the credit recovery through crop payment is low because farmers sell on the open market to evade repayment of their government loan. Therefore, food shortage is buffered mainly through rice

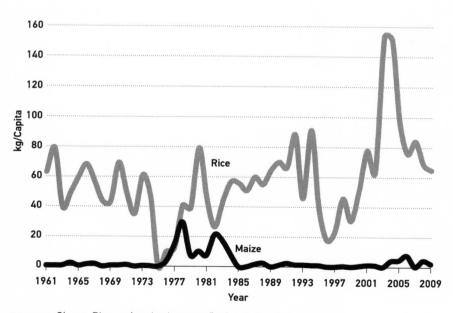

FIGURE 5. Ghana: Rice and maize imports (kg/capita), 1961 to 2009. SOURCE: FAOSTAT.

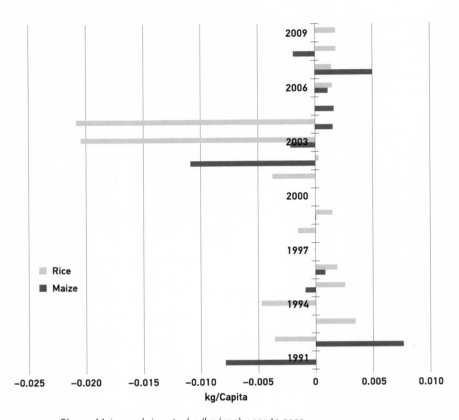

FIGURE 6. Ghana: Maize and rice stocks (kg/cap), 1991 to 2009. SOURCE: FAOSTAT.

importation; the first surge in rice importation was in the early 1980s when Ghana was hit by a severe drought. The surge in more recent years, which hit the ceiling about 2005, can be attributed to an increase in consumer demand for imported rice. This demand was fuelled by government import subsidies on the commodity and by commercial advertisements by importers such as Chicago Star Rice.[5]

Impact of Policy Programs on Priority Crops

The impact of the policy programs on the various priority crops can be assessed by comparing consumer prices among the priority crops. Food import has the effect of holding the consumer price of the imported commodity at low levels. Input subsidies also drive down consumer prices indirectly through artificially reducing private production costs.

Among the priority crops for which price data are available, the price of maize is the lowest (figure 7).[6] Although importation levels of the crop and available levels of subsidized fertilizer and tractor mechanization are low, maize benefits from the use of mechanical shelling machines. The use of the shelling machines eliminates labor input in maize processing; the crop is marketed in grain form (Ghana 2010). The impact on cassava cost reduction in Nigeria of using a mechanical grater for processing *gari* suggests that the use of the mechanical shelling machines could have a considerable impact in terms of reducing maize production and processing costs. The use of the mechanical grater in processing *gari* reduced the

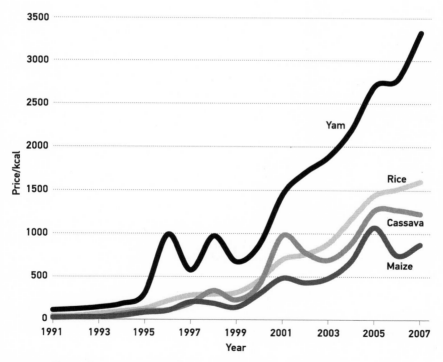

FIGURE 7. Ghana: Consumer prices of yam, rice, cassava, and maize (GHC at current prices), 1991 to 2007. SOURCE: FAOSTAT.

gari processing cost by 50 percent. Grating is one of the cassava processing tasks; others are peeling and toasting.

Several NGOs and government parastatals are promoting maize in Ghana. For example, Masara N'Arziki (meaning Maize for Prosperity) is an NGO that is headquartered in Tamale in the Northern region of Ghana but covers the entire northern part of the country. The Masara N'Arziki Farmers Association is part of the Industrial Maize Program established in 2005 by Wienco (Ghana) Limited to support maize farmers through the supply of improved technologies to increase productivity. The program package consists of fertilizers, hybrid seeds, herbicides, insecticides, spraying equipment, improved farm implements, and technical advisory and training services, provided to farmers on credit.[7] Other agencies that promote maize in the country include the Savannah Accelerated Development Authority, an independent government agency; the Adventist Development and Relief Agency; and World Food Program.[8] None of the NGOs or parastatals promote yam.

Cassava is not imported in Ghana, and the crop does not benefit from most of the input subsidy programs except the initial free distribution to farmers of the planting setts of the IITA's mosaic resistant, high-yielding TMS varieties in the early 1990s. After this, farmers are comfortable using planting setts from their fields and they usually do not apply fertilizer to cassava. The low price of cassava reflects its low production costs relative to yam and availability of mechanical laborsaving technologies for processing, such as the grater for preparing *gari* and the mechanical mills for converting dried roots into flour.

The impact of input subsidies on the price of rice is low because of the low availability levels of the subsidized inputs. The lower consumer price of rice relative to yam reflects the impact of large amounts of rice importation on the domestic consumer price of the commodity.

The consumer price of yam is highest, by a large margin, among the four priority crops for which price data are available. Yam is not imported in Ghana; rather it is exported, and among the priority crops yam benefits the least from the input subsidies because of low production and postharvest technologies in the yam food sector. Yam seed technology is low; there are no high-yielding yam varieties at the farm level; production and harvesting of yam are not mechanized; and yam storage technology is rudimentary. Yam is not part of the credit subsidy program extended to farmer groups because of the rudimentary storage technology. Loan recovery through crop purchase and stockpiling is not feasible in the case of yam because the crop cannot be stored on a large scale.

Summary

Low production and postharvest handling technologies place yam at the periphery of Ghana's national food policy programs, which are based on improved technologies. The result is a high consumer price for yam, which is higher by a large margin than the price of any of the other priority crops. The implication is that farmers' adoption of available new technologies such as yam minisett and hybrid yam varieties, as well as development and diffusion of other essential technologies, such as storage technologies, will help position yam at the center of national food policy programs. This will in turn help to drive down consumer prices of yam. The problem of yam pests and diseases is an impediment to the diffusion of available technologies. Therefore, control of yam pests and diseases will help bring yam into focus in national food policy programs and permit the crop to benefit from public investments in the development of the yam food subsector.

Yam slices used as seed for producing seed yam by minisett technique.

Seed yams for producing ware yams.

Village carvers fashion hand hoe handle for yam production in Zaki Biam, middle belt of Nigeria, January 2013. COURTESY: THE AUTHOR.

Village smiths fashion hand hoe blade for yam production in Zaki Biam, middle belt of Nigeria, January 2013. COURTESY: THE AUTHOR.

Aeroponics seed yam system, IITA, Ibadan. COURTESY: NORBERT MAROYA.

Yam mound making, Abakaliki, eastern Nigeria. COURTESY: LOUISE FRESCO.

Roasting yam in open fire along highway for sale to travelers. COURTESY: THE AUTHOR.

Ware yam market scene, Zaki Biam, Nigeria. COURTESY: THE AUTHOR.

Ware yam market scene, Kintamkpo, Ghana.

Women arrive at a wedding reception bearing gifts in Ukpo, Dunukofia, eastern Nigeria, September 2010.

Community elders assemble for New Yam festival in Ukpo, Dunukofia, eastern Nigeria, September 2010. COURTESY: JEFFRY OLIVER.

The author (*right*) with his professional guide of over fifty years, Professor Carl Eicher (*left*) at Nsukka, eastern Nigeria, 1986. COURTESY: NWOGO NWEKE.

Dr Regina Kapinga

Prof Felix Nweke
BMGF

The author (*seated*) at Yam Improvement for Income and Food Security in West Africa (YIIFSWA) workshop in Accra, Ghana in 2012. COURTESY: OIWOJA ODIHI.

CHAPTER 9

Yam Stereotype as a Man's Crop

This chapter relies on farm-level data generated by the COSCA studies in Nigeria, Ghana, and Côte d'Ivoire. The sample for the COSCA studies was focused on areas where cassava is grown, but the West African yam belt is a subset of these areas. The COSCA field-level data were collected in the early 1990s, but an extensive literature search on yam conducted for this book failed to turn up evidence of an update in the rich COSCA database. Besides, most, if not all, yam production and postharvest technologies have not changed in living memory.

A Historical Perspective of Gender in African Agriculture

In Africa, one of the first to recognize the importance of women in farming was Hermann Baumann in 1928, with his article "The Division of Work According to Sex in African Hoe Culture" (Baumann 1928). Phyllis Mary Kaberry published another important study of women in the Cameroon in 1952, and R. Galletti, K. D. S. Baldwin, and I. O. Dina published empirical data on male and female activities in the book *Nigerian Cocoa Farmers* in 1956 (Kaberry 1952; Galletti, Baldwin, and Dina 1956).

Ester Boserup, a Danish social scientist, provided an array of evidence to show that women in developing countries play significant roles in agricultural and rural development and that Africa was the region of female farming par excellence. Her monograph *Women's Role in Economic Development*, published in 1970, gave birth to the Women in Development (WID) approach that emerged in the 1970s, calling for treatment of women's issues in projects. In the monograph, Boserup drew world attention to what she understood to be the role and plight of women in African agriculture. She reported that, in many African tribes, nearly all the tasks connected with food production were carried out by women. Boserup drew on eighteen anthropological village studies and concluded that women in Africa often "do more than half of the agricultural work; in some cases they were found to do around 70 percent, and in one case nearly 80 percent of the total" (Boserup 1970, 22).

Later on, in the 1980s, an alternative approach—the Gender and Development (GAD) approach—proposed more emphasis on gender relations in general. The approach is more concerned with relationships, the way in which men and women participate in development processes, rather than strictly focusing on women's issues (Cook and Curran 2009). While WID and GAD approaches brought gender relation issues, such as the roles of men and women in agriculture, resource control by men and women, and impact of development actions on men and women, to the forefront, the labeling of a crop as man's or woman's is a further dimension of the subject of gender in African agriculture.

In a 1982 study of a root crop farming system in eastern Nigeria, E. C. Okorji found that some root crop fields were owned separately by men and women who were sometimes members of the same households, while other fields were owned together by the household. Within each field, irrespective of gender ownership, yam was considered a man's crop and cassava a woman's crop, though the crops were grown as intercrops in the same field at the same time (Okorji 1983). Yam was also considered a man's crop and cassava a woman's crop in fields cultivated by women heads of households that had no adult males, as well as in fields cultivated by men in households with no adult females. Okorji concluded that crop labels as men's or women's in eastern Nigeria were stereotypes (Okorji 1983).

The origin of stereotyping yam as man's crop in West Africa can be speculated. First, in several West African cultures, wealth was controlled by the man who served as head of the household, and in the past yam was the ultimate wealth because it was the major crop where agriculture was the main business. Cocoyam was a minor crop, and it was labeled a woman's crop. Cassava, maize, and rice came later in the history of the region. Second, some of the yam production activities such as cutting through forests to bring new land into cultivation and making mounds for seedbed are labor intensive and require muscular men to accomplish.

D. C. Ohadike reported that the yam production requirement of masculine labor was a contributing factor to the expansion of cassava production in the Lower Niger region of Nigeria at the turn of the twentieth century.[1] In the Lower Niger region, a series of three tragedies—a war of resistance against the imposition of British rule (1899–1914), the First World War (1914–18), and the influenza epidemic (1918)—made sustenance through yam production difficult. Yam production was adversely affected by the withdrawal of men from the villages to fight in the wars. Consequently, the people of the Lower Niger region embraced cassava, which had been hitherto considered inferior to yam but was less labor intensive to produce (Ohadike 1981; Chiwona-Karltun 2001).

Gender Roles in Yam Production and Postharvest Activities in Nigeria, Ghana, and Côte d'Ivoire

Since yam is stereotyped as a man's crop, should technologies be channeled through men or are there situations where technologies are more efficiently channeled through women? To answer this question, the roles of men and women in terms of yam field ownership and contributions to various yam production and postharvest activities will be evaluated to define the best channel of delivery of different yam technologies. This is of interest because of the considerable investment being made in the development of a wide range of yam technologies in West Africa at this time.

In Nigeria, Ghana, and Côte d'Ivoire, COSCA researchers took a sample of households in several representative cassava-producing villages. A household cultivated multiple fields in which they practiced intercropping. Farm-level information was collected firsthand on all fields cultivated by each of the sampled households. In each field, the head of the household, usually a man, and his wife were asked who owned each crop in the field, a man, a woman, or a combination of household members. The couple was also asked who performed most of the farm tasks in the field—namely, land clearing, seedbed preparation, sowing of each crop, weeding, harvesting, and transporting of each crop from the field. Marketing information was collected in the market survey component of the yam consumption study and will be discussed later.

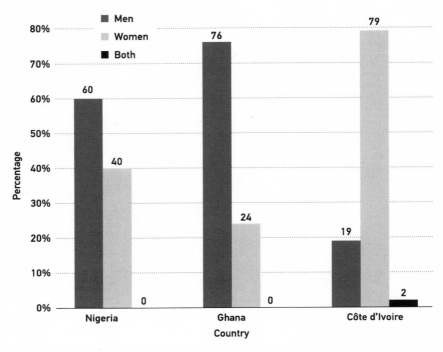

FIGURE 8. Nigeria, Ghana, and Côte d'Ivoire: Percentage distribution of yam field ownership by gender, 1991. SOURCE: COSCA DATA.

Analyses of the COSCA data on yam field ownership by gender disclosed that in Nigeria, Ghana, and Côte d'Ivoire men and women owned yam fields, that is, both the male and the female genders grew yam in their own rights and made production, marketing, and utilization decisions. In Côte d'Ivoire, women owned more yam fields than men, while in Nigeria and Ghana men owned more than women (figure 8). Matrilineal and patrilineal societies are both found in West Africa, but matrilineal societies are more common in Côte d'Ivoire than in Nigeria or Ghana. In the past, family stock of seed yam was used as a means of transferring family wealth from one generation to the next. In matrilineal societies women inherit the stock of family seed yam, while in patrilineal societies the stock of family seed yam is inherited by men. In Côte d'Ivoire, women own more yam fields than men because matrilineal societies are more common in that country than in Nigeria and in Ghana.

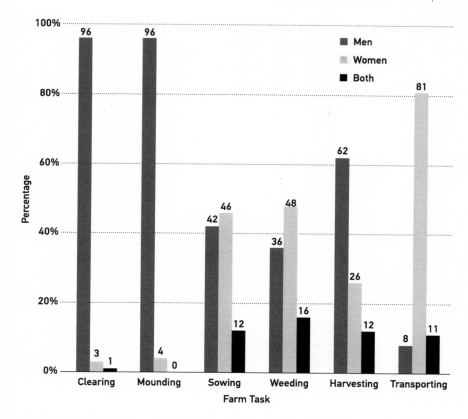

FIGURE 9. Nigeria, Ghana, and Côte d'Ivoire: Percentage of yam field tasks performed by men and women, 1991. SOURCE: COSCA DATA.

The analysis of the COSCA data further reveals that in each of the three countries the number of yam fields in which women provided the bulk of the labor for each task increased from a low level during land clearing to a higher level at weeding and transporting (figure 9). By contrast, the number of fields in which men provided the bulk of the labor was highest during land clearing, mound making, and harvesting. These findings show that *both* men and women are heavily engaged in different yam production and postharvest tasks in the three most important yam-producing countries in the world. A different gender study in the agricultural systems of the Niger Delta region of Nigeria revealed a less rigid division between male and female roles for traditional crops, such as yams (in the past, male responsibility) and cassava (female responsibility), where men and women are now cultivating both crops (IITA 2004).

These studies contradict Boserup's conclusion of forty-five years ago that women in Africa often did more than half of the agricultural work (Boserup 1970). Yet the near monopoly of women in yam marketing stands out. The yam consumption study revealed that in the Techima yam market in Ghana, more than 90 percent of the wholesale traders were women; these were not just errand girls, but wholesalers with their own stalls. In fact, in that market, all members of the yam traders' union executive committee from the chairperson to the last member were women. These women were calling all the shots in the market. In the Accra yam market also in Ghana, the story was similar; there the few men wholesale traders, whom the yam consumption study researchers met, would not grant an interview until they obtained the consent of the female chairperson of the union executive committee. In each market, women were also retailers and young men were the carriers who transported yam between trucks and stalls by wheelbarrows. The young men loaded and unloaded yams from trucks for payment by the women traders. In Nigerian yam markets, women wholesale traders still outnumbered men (6:4) though at a lower margin than in Ghana (9:1).

The division of labor by task in the yam food sector is influenced by a number of factors, including the power intensity of the different tasks that are performed (Spencer 1976). For example, manual land clearing and mound making for yam planting are power-intensive tasks that are usually performed by men, but in some isolated cases women also perform some of these tasks. In the case of cassava, a COSCA researcher in the Congo asked a male farmer who—men or women—provided the bulk of labor for transporting cassava from his field. The farmer retorted that the COSCA investigator should know that only women could perform the task because transporting cassava by carrying on their heads or backs was a very heavy task.

In each country, the trends in the level of contribution of labor by gender to the yam production and postharvest activities are such that as the activities move from the field toward home, women's contributions increase and men's contributions decline. Why? Women do more of the work that is close to home than men to accommodate homemaking activities such as preparing meals and caring for children (Cook and Curran 2009).

To summarize, although yam is stereotyped as a man's crop, men and women produce yam in their own rights, which permits each gender to make production and postproduction decisions. Both men and women engage in varying but complementary yam production and postproduction tasks. Men more frequently perform tasks away from the home, such as land clearing and mound making. Women more frequently perform tasks closer to home, such as transportation and marketing, which they are able to combine with homemaking activities. In yam technology development and transfer, these roles of men and women are more pertinent issues for consideration than the labeling of yam as a man's crop.

Consequences of the Gender Roles for Yam Technology Development and Transfer

Mound making creates the first labor bottleneck in yam production. A change in technology that eliminates or reduces that bottleneck will release labor, especially male labor, from mound making for use in the next highest labor demanding activity. That activity is weeding, which is presently carried out mostly by women. Having men work more on weeding will lead to expansion in yam production.

A farmer group interviewed in the Tamale area of northern Ghana, most of whom were export yam producers, reported that they narrowed the number of yam varieties they grew to the few bought by traders. This underscores the influence that women, who have a near monopoly of yam marketing in the country, have on the choice of yam varieties produced in Ghana. The implication is that new yam varieties could be more easily delivered to farmers through yam wholesale traders who are mostly women.

Boserup argued that "women usually lose in the development process" because men monopolized the use of new equipment and agricultural methods and that this tendency was frequently reinforced by a bias in extension programs in favor of men. As a result, there will be a relative decline in the productivity of women and "the corollary of the relative decline in women's labor productivity is a decline in their relative status" (Boserup 1970, 53). In Nigeria and Ghana, would women lose in a process of development in yam marketing?

Summary

In many parts of West Africa, yam is referred to as a man's crop, but that is only a stereotype; the rich database generated by the COSCA studies shows that men and women produce yam in their own rights. Both men and women engage in varying but complementary yam production and postproduction tasks. Men more frequently perform tasks away from home, such as bush clearing and mound making, and women more frequently perform tasks closer to home, such as transportation and marketing.

A change in technology that breaks the yam mound-making labor bottleneck by eliminating it, or at least reducing it significantly, will release labor, especially male labor, from that activity for use in the next highest labor-demanding activity, namely weeding, and thereby lead to expansion in yam production. New yam varieties could be more easily delivered to farmers through yam wholesale traders. These yam wholesale traders are mostly women who have considerable influence on the choice of yam varieties produced in Ghana.

Ester Boserup's conclusion that women in Africa often "do more than half of the agricultural work, in some cases they were found to do around 70 percent, and in one case nearly 80 percent of the total," is confirmed in the particular case of yam marketing in West Africa (Boserup 1970, 22). Her argument that "women usually lose in the development process" pleads for investigation in the case of the yam marketing in West Africa (Boserup 1970, 53). Does women's virtual monopoly of yam marketing in West Africa hurt them, in what ways, and to what extent? An in-depth yam marketing margins study in West Africa would shed some light on this question. The study should go further to characterize the women yam wholesalers in terms of age, experience, and time management. This will help determine how much of their time is invested in the yam trade and how much is left for their other responsibilities, especially caring for children. Where applicable, in-depth yam market analyses will allow for a comprehensive overview of the private and social benefits or costs of the dominance of women in yam marketing in West Africa.

CHAPTER 10

Yam Postharvest Matters

In this chapter, storage, marketing, and prospects for the processing of yam are discussed as distinct yam postharvest matters. Storage is an on-farm postharvest function because yam is stored on the farm. Transportation from the farm to selling points, such as wholesale and retail market centers, is more significant than on-farm transportation. Processing is discussed in terms of its prospects because at present that postharvest activity is low in the yam food sector in West Africa.

On-Farm Storage

On-farm yam storage technologies vary from place to place depending on a range of circumstances, including security and ecological conditions. Common storage methods in yam-producing centers surveyed for the yam consumption study in Nigeria and Ghana are piling in thatched mud huts with perforated walls, tying on racks, and piling on the ground in dry material-covered heaps. All the existing storage techniques date back to a time not remembered by anyone in any of the farmer groups interviewed in the survey, even the oldest farmer in the area (Nweke, Aidoo, and Okoye 2013). Therefore, yam storage methods vary in space but not in

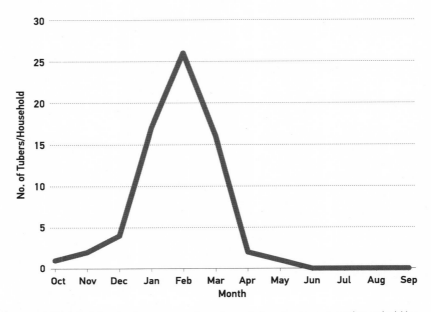

FIGURE 10. Eastern Nigeria: Average number of yam tubers in storage per household by month. SOURCE: UGWU 1990.

time; that is, different yam-producing areas have different methods that have not changed over time.

Each existing yam storage method has an upside and a downside. Security and enhanced aeration, especially in high humidity environments, are the main advantages of storage by tying on racks at home, while the high labor requirement is the major drawback of that method. In less humid environments, storage in a thatched hut or in a covered heap is preferred because of the minimal amount of labor need. Both of these methods expose yam to pests and diseases. In a study of resource productivity in yam-producing areas of southeast Nigeria Dr. Boni Ugwu of the Nigerian NRCRI took a course route approach to measure net quantity of yam brought in less quantity taken out of store among farmers in eastern Nigeria (Ugwu 1990). Dr. Ugwu's data reveal that stock began to accumulate in October, peaked in February, and thereafter diminished to zero in June (figure 10).[1]

In eastern Nigeria, harvesting of yam starts in June, and the peak period is from November to February. Yams harvested early—that is, before November—are

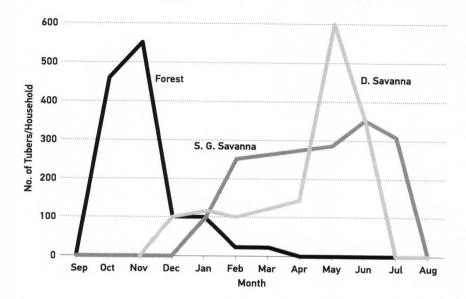

FIGURE 11. Eastern Nigeria: Average monthly yam sales (no. of tubers/per household) from storage by ecological zone. SOURCE: UGWU 1990.

usually used or sold immediately. As a result of high water content, yams harvested before November are more susceptible to damage while in storage by pests and diseases than yams harvested later in the season. However, ecological differences in planting and harvesting dates guarantee year-round availability of yam, with a peak around May, even though storage technologies are rudimentary. Dr. Ugwu's data reveal that stored yams from the forest zone are in the market mainly between October and January; from the derived savanna, December to June; and from the Guinea savanna, January to July (figure 11).

Each of the graphs in figure 11 steeply drops, to a low level or to zero; at a certain stage in the storage process, yam tubers are withdrawn from storage and used or sold to minimize weight loss due to sprouting and dehydration or total loss due to pest and disease problems. Yam has high value per unit weight; whenever possible yam tubers are not allowed to be wasted. None of the existing storage techniques is capable of holding yam from one harvest to another.

The result of existing storage technologies' inability to hold yam from one

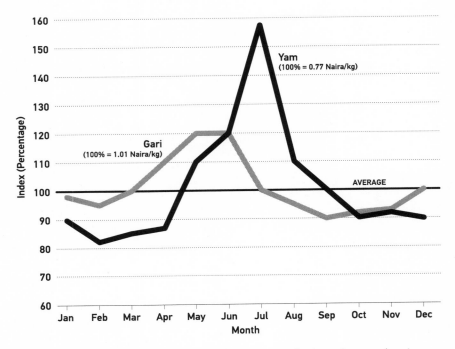

FIGURE 12. Eastern Nigeria: Indices of monthly average retail prices of yam and *gari* (cassava product). SOURCE: UGWU 1990.

harvest to another is supply and market price seasonality. In Africa, food crops are used more at harvest than at other times of the year, mainly because of higher availability and consequently lower market prices. An additional reason, in the case of yam, is fear of damage; the rate of damage in stored yam by pests and diseases is highest for tubers with physical injury, which are often sustained at harvest. Figure 12 compares indices of average monthly retail market prices in Eastern Nigeria of yam with that of cassava. The coefficient of variation is considerably higher in the case of yam, 7.83, than cassava, 1.30.

These scenarios imply that famers incur capital losses because they are compelled by fear of damage to sell most of their yams at low prices at harvest and that consumers pay high prices later in the season when supply is low. Here, the takeaway is that improvement in yam storage technologies will improve producer income and stabilize prices paid by consumers.

Issues in Yam Marketing

Marketing, the business activities engaged in the flow of goods and services from producers to consumers, bridges the gap between the two functionaries. The marketing system is composed of marketing channels that, in the case of agricultural marketing, have links as rural assembly markets, wholesale markets, and retail markets, each providing a different marketing function (Ezedinma et al. 2007). Ordinarily, rural assembly traders should buy from farmers with whom they are in close proximity and sell to wholesale traders in distant urban market centers. With this arrangement, farmers would be able to concentrate on the business of production. In the urban market centers, wholesale traders should perform a speculative function in warehouse storage to even out supplies from times of surplus to lean periods. How efficient are these marketing functions in the case of yam in West Africa?

Dearth of information is a major issue in yam marketing in West Africa. There are few reports of yam marketing studies. The marketing component of the yam consumption study found out with dismay that there was no record of yam traded or of yam prices in the major yam markets in both Ghana and Nigeria. Yet simple records of yam movements in volume and value will be a powerful tool for establishing the economic importance of yam in each country. It will be an effective tool of advocacy among agricultural policy makers and donor organizations for investment in yam R&D.

In most of the West African yam belt, the yam marketing channel is weak in terms of rural assembly and wholesale warehousing. Most yams that are for sale move from the farm to consumers as quickly as possible, with limited rural assembly and wholesale speculative functions, for fear of losses from pests and diseases (Amikuzuno 2001). In the absence of sufficient rural assembly traders, how does yam move from the farm to the wholesale market? The market survey component of the yam consumption study revealed that in some cases, urban wholesale traders bought yam at the farms, and in other cases farmers sold yam themselves at the urban wholesale markets.

When an urban wholesale trader buys at the farm, he discounts the transportation cost and the cost of his time from the price he offers. When a farmer sells at the urban wholesale market he figures in transportation cost and the cost of his time in the price he asks for. Bargaining power determines who gets the price he asks; the farmer who paid transportation costs to get from the farm to the distant

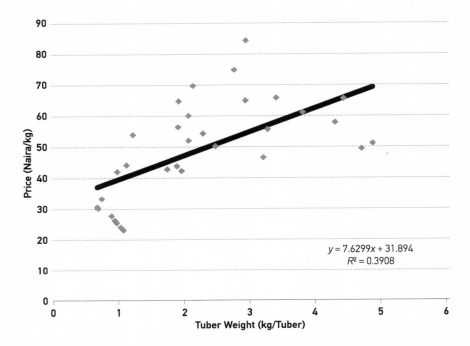

FIGURE 13. Nigeria: Wholesale price of yam by tuber weight in Shaki market, May 2013.
SOURCE: NWEKE, AIDOO, AND OKOYE. 2013.

urban wholesale market center with a truckload of yam, which is perishable, is in a weaker bargaining position because the cost to the trader is lower as he arrives at the farm empty handed.

Therefore the limited rural assembly function in the yam marketing channel, which is caused by the fear of pests and diseases, does not favor yam producers. Similarly, the limited speculative function in the yam wholesale market that is also caused by the fear of pests and diseases does not favor consumers who pay high prices during lean periods. In West Africa, the dread of pests and diseases is as much a problem in yam marketing as it is in yam production.

How are the prices set? This question was posed to the wholesale traders interviewed in the market component of the yam consumption study. Price setting was simple: farmers ask for a price, buying wholesalers negotiate downward, and agreement may or may not be reached. The same selling process is followed by wholesalers selling to retailers. The yam consumption study researchers weighed samples of yam tubers with a scale. They were weighed in several heaps of varying tuber sizes in the Shaki yam market in Nigeria and the Techima yam market in Ghana. This painstaking exercise revealed that price per unit weight is higher for heaps of large tubers; that is, the larger the tuber, the higher the price per unit weight (figure 13).

IITA's YIIFSWA baseline survey data revealed that, in Ghana, there is a problem at the level of outlet for yam from the farm to urban markets.[2] A yam producer faces three options: one, the producer takes out a loan from a trader at planting time to pay the high production costs, and at harvest the trader arrives at the farm with a truck and carries the yam away at his price; two, a trader also arrives at the farm of a producer who did not receive a loan and makes an offer, and if the price is not acceptable to the producer the trader goes to another farm with his empty truck; and three, the farmer takes a truckload of his yam directly to the urban market and stops at the entrance (the league of middlemen prohibits farmers from entering the wholesale market with yam) where a middleman takes the truckload, negotiates a price with the wholesalers behind the farmer's back, and takes his agreed and unagreed commission before handing the proceeds over to the farmer who does not know what the wholesaler paid. Unagreed commission would be part of the proceeds that the middlemen often take beyond the agreed commission since the farmer is unaware of the amount the wholesale buyer paid for his yam.

Obviously, the context in which yam producers sell their crop has a high potential to impoverish them; policy interventions are needed to change the unfair situation. The first step is to empirically assess the marketing situation to determine if the yam traders are enriched by the context that impoverishes the farmers. The empirical assessment will identify measures that, if implemented, will enable all participants in the yam value chain, the producers as well as the traders, to be equitably compensated for their efforts. The present situation is undesirable because consumers pay high prices and producers live in poverty.

An improved transportation system in Nigeria and Ghana is powerful in driving the yam marketing system. Within the limit of some serious impediments, the transportation system has proved to be effective in moving large quantities of yam on a daily basis from producing areas to urban market centers in each country. The transportation system is mechanized with motorized vehicles, especially from the farm to distant wholesale urban markets, and in each country parts of the intercity roads are paved. The transportation function remains expensive because of a wide range of impediments, including bad roads in some sections, high vehicle running costs, and fraudulent security and customs agents along the roads. The high transportation costs combine with high production costs to drive up the price of yam to consumers.

In Nigeria and Ghana, there is a designated yam market in most of the yam-producing zones principally for the wholesale of yam. In Nigeria, there are yam markets in Zaki Biam, Otuocha, Shaki, and Iseyn among others. In Ghana, yam markets are in Accra, Kumasi, Techima, and Kintamkpo. In yam markets, yams are sold in heaps of 100 tubers of approximately uniform size in Nigeria and 120 in Ghana; the tuber size is determined by handling and appearance. The heaps are piled out in the open, or under thin metal roofs under which conditions the yams are exposed to heat.

Each yam market is governed by a yam traders' union–elected executive committee; the governance in each market is seller oriented and does not address buyer interests, such as the absence of standards and, the most problematic, the absence of records. In each yam market the union executive has an office and collects taxes from buyers and sellers but does not keep any records. The result is that the amount of yams traded, prices paid, etc., are unknown.

Odds on Yam Processing and New Uses

In Nigeria and Ghana, yam is used as food; use as feed or industrial raw material is minimal.[3] The reason for this is the high cost of yam relative to alternative crop sources, such as cassava as well as corn, which is imported from Europe and North America. In terms of feed and industrial raw material, the value of yam, cassava, or corn, respectively, is in its starch content, which is high in all three commodities. The potential for using yam as feed or industrial raw material in Nigeria and Ghana depends on its price relative to the prices of cassava and imported corn. Presently, the relative price is in favor of imported corn first, followed by cassava. Therefore the likelihood of yam being used as feed or industrial raw material in Nigeria and Ghana depends on driving down its production costs to make its price competitive with the prices of cassava and imported corn in the feed and industrial raw materials markets.

Yam faces the same challenge of high costs in food industries as in feed and industrial raw material markets. Often at yam conferences, workshops, and related gatherings, food scientists excitedly display pastry products made from yam flour.[4] Such excitements are justified because the products are proof that making pastries from yam flour is technically feasible, but economic feasibility needs to also be established to consummate the excitement. That can be done by bringing the cost of yam down to the cost levels of cassava and imported grains. In West Africa imported grains are cheaper than yam and cassava; for this reason, pastries are made mostly with grain flour, especially wheat flour.

In West Africa, the consumer preference of yam is for fresh tubers instead of a preprocessed form. Yam food is prepared into any number of food forms from fresh tuber when needed, including boiled yam (called yam *ampesi* in Ghana), fried yam (i.e., yam boiled in oil), yam roasted or grilled on an open fire, and pounded yam, also called *foofoo, fufu, or foutou*. At present, the processed form of yam is a dried tuber that is milled into flour for preparation as *foofoo*. This means that the presently available processed yam has a limited value only as a substitute for fresh tuber *foofoo*; yam consumed in any other food form is prepared from fresh tuber.

In Nigeria, dried yam tuber flour is processed in one of two ways: *poundo* yam prepared industrially and *amala* prepared in the traditional sector. *Amala* is a preferred yam food product to some consumers in the traditional sector that is within parts of western Nigeria. *Poundo* yam is consumed as a matter of convenience rather than preference because consumer preference is for pounded yam.

Poundo yam has alternatives—namely, *semo* (a form of corn or wheat flour) and dried cassava root flour—that are also prepared as *foofoo*. Traders interviewed for the yam consumption study in the Otuocha yam-producing area of eastern Nigeria reported that *poundo* yam is not popular in the area. Traders do not like to stock it because it is more susceptible to pests and costs more than its substitutes.

Two conclusions emerge from these analyses; one is that the odds of using yam as feed or industrial raw material depend on implementation of measures that will drive down the cost of yam relative to alternative crop sources of feed and industrial raw material. The other conclusion is that increased consumption of yam in preprocessed forms depends on the development and dissemination of technologies for processing boiled, fried, or roasted yam, improving the quality and reducing the cost of *poundo* yam to make it competitive with pounded yam and its substitutes.

Summary

The inability of existing yam storage technologies to hold yam from one harvest to another creates seasonality in supply and in market prices, leading to more use at harvest than at other times. The shortage of market information, dread of yam pests and diseases, reactionary pricing strategies, and high transportation costs are principal issues of concern in yam marketing in Nigeria and Ghana. Apart from these, other deficiencies that need fixing include the lack of standards and measures and the practice of displaying yams in the open under sun and heat.

The yam marketing channel is short in terms of rural assembly and wholesale warehousing because of the fear of pests and diseases by the traders. This situation works against yam producers who have to combine production and marketing functions. It also works against consumers who pay high prices during lean periods, creating the undesirable situation of yam producers living in poverty while consumers pay high prices. The dread of pests and diseases, as in yam production, is one of the bottlenecks in yam marketing in West Africa.

The prospect for using yam as feed and industrial raw material depends on the implementation of measures that will drive down the cost of yam relative to alternative crop sources of feed and industrial raw material. An increase in the consumption of yam in preprocessed forms depends on development and dissemination of technologies for improving the quality of *poundo* yam to make it quality competitive with pounded yam and reducing its costs to make its price competitive with substitutes.

CHAPTER 11

Trade in West African Yam

Recorded information is as scanty and unreliable on yam trade as it is on yam marketing in West Africa. FAOSTAT records yam exportation from Ghana and yam importation in Mali; records of exportation or importation in Nigeria and Burkina Faso are scanty. Yam produced in West Africa is widely available in Europe, in North and South America, and in parts of Asia where the food crop is consumed mostly by growing West African immigrant populations. This chapter relies on the yam consumption study information collected in two major yam-producing countries—Nigeria and Ghana. Also important is the information collected in two minor producing countries in West Africa—Burkina Faso and Mali—as well as observations made in extensive interstate road travels to observe yam movement across the porous borders in West Africa. Because of the lack of reliable quantitative data, no attempt is made to estimate present volumes or values of trade in this commodity, which is rapidly emerging as a major West African export crop.

Evidence of Trade in Yam within the West African Region

On December 4, 2012, at the Sankare Yaare market in Ouagadougou, Burkina Faso, large quantities of yams were on display for sale. They were mostly white yams. Upon interview, the sellers reported that the yams were entirely imported from Ghana and that yams from anywhere else faced low consumer acceptance in the country. Similarly, on December 7, 2012, at the Soukouni Coura de Medine market in central Bamako, large quantities of yams were also on display for sale. In contrast to Sankare Yaare market in Ouagadougou, the yams displayed in Soukouni Coura de Medine were mostly water yam, which the traders disclosed were mostly imported from Côte d'Ivoire.

In each major producing country, such as Nigeria and Ghana, yam is shipped from many markets and arrives in minor producing countries such as Burkina Faso and Mali at many different entry points without records of the movement in terms of frequency, volume, or value. The bus travels for the yam consumption study revealed that, despite custom checks at various official border posts, baskets of yams were taken across by bus riders and by head loads carried by pedestrians.

National Government Trade Policies Relevant to Yam

Standard intervention measures in trade such as tax breaks and export/import subsidies are administratively difficult to implement by poorly performing governments such as those of Nigeria, Ghana, Burkina Faso, and Mali. In Ghana, the government is attempting to develop and implement policies, such as tax rebates, custom duty exemptions, and simplification of export procedures, to benefit nontraditional export commodities, which include yam, through the Ghana Export Promotion Council (Thrupp et al. 2006).

Yam trade among West African countries is being driven more effectively by regional policies such as the policies of the Economic Community of West African States (ECOWAS) than by individual exporting or importing country policies. Three ECOWAS policies that are aimed at trade liberalization in the region and favor yam trade come to mind: the development of interstate roads linking countries in the region, the free movement of goods with low added value, and the free movement of people without visa restrictions. The consumer preference for white yam or water yam may not provide a full explanation for the movement of yam from Ghana and

Côte d'Ivoire to Burkina Faso and Mali, respectively, because Ghana produces water yam just as Côte d'Ivoire does. A more likely explanation could be in differentials in transportation costs. Movement of yam from Kumasi in central Ghana to Bamako by road is through Burkina Faso, while movement from Côte d'Ivoire is direct to Mali.

In West Africa, a program of interstate highway construction is ongoing; each West African country is required to construct the road within its borders. Interstate bus travels for this study from Lagos through Cotonou to Lomé, Ouagadougou, Bamako, Kumasi, Accra, and back to Lagos revealed that this program is advancing in all the countries passed through, namely Nigeria, Benin Republic, Togo, Burkina Faso, Mali, and Ghana. Some stretches of the roads can be described as fair to good and some are under repair. Other stretches are in such poor condition that the high cost of wear and tear on transport vehicles is palpable and can easily be seen as a factor that is driving up the cost of yam through high transportation costs.

For example, survey travel by bus from Bobo in Burkina Faso to Ghana was on a very poor road. On the way to Wa in Ghana the bus had to have a broken axle repaired. After close to five hours of waiting, another bus took the passengers to Kumasi. This experience suggests that yam importation from Ghana to Burkina Faso through the same road is at a high cost of transportation. Continued improvement of these interstate highways will certainly help boost yam consumption by driving down transportation costs and, eventually, the price of yam to consumers in importing countries.

For some time now, citizens of ECOWAS countries have traveled freely among member countries without a visa or passport; only a West African travel certificate, which is easier and cheaper to acquire than a regular passport, is required. To complement this policy in terms of yam trade is the policy of free movement of goods with low added value.[1] Both policies are in favor of yam. Free movement of people allows traders and individual consumers to move yam in small quantities across borders. The policy of free movement of goods with low added value is particularly in favor of yam, which is traded at its fresh form and therefore has zero added value.

Obviously, yam exportation to countries outside West Africa depends on demographic movements. Governments' policies of relaxation of emigration laws are likely to favor yam exportation outside West Africa. For example, the Nigerian government's dual citizenship policy makes it easier for Nigerian citizens to reside in other countries, such as the United States, that have a similar policy.

To summarize, volumes and values of yam trade in West Africa are unknown because of a lack of recorded data. West African yam is traded globally. This is

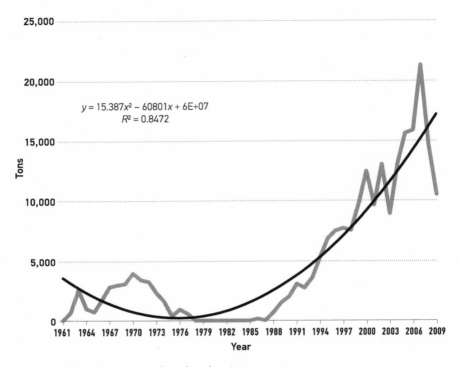

FIGURE 14. Ghana: Yam export (tons/year), 1961 to 2009. SOURCE: FAOSTAT.

driven by growing African populations in diaspora. Regional trade between major and minor producing countries within West Africa is driven by growing populations and incomes in importing countries, such as Burkina Faso and Mali, and by regional ECOWAS policies of development of interstate highways and liberalization of immigration procedures.

Yam Trade between Ghana and Burkina Faso

In the past fifty years, Ghanaian yam exportation has had a strong positive trend; beginning in 1962, soon after Ghana gained independence in 1957, yam exportation increased rapidly (Figure 14).[2] But the trend sagged to zero in the mid-1970s through the 1980s following economic, political, and climate crises the country experienced during the period. After the crises, the trend surged again to high levels, but seemed to be slowing down again beginning around 2008, perhaps because of climate change, declining soil fertility, or the increasing buildup of yam pests and diseases. Population growth in the country and redundant production and postharvest technologies in the yam food sector could also be reasons for this decline.

A future trend in exportation defined by production surplus over consumption need in Ghana is projected. The future is limited to 2025 because of the low prediction ability of the projection procedure (Nweke, Aidoo, and Okoye 2013). Past production trends are assumed for the production projection, and population, GDP per capita growth rates, and 0.05 income elasticity of demand for yam are the only bases for consumption projections. The projections are based on a procedure that is limited in robustness and on data that have doubtful credibility; the value of the projections is as an indicator of what the future level of exportation can be.

FAOSTAT is blank for yam in the seed column; but in 2009, FAOSTAT allocated 35 percent of gross production to "Other Util" in Ghana. Projected gross production is discounted for seed by 35 percent to arrive at the gap between production and consumption. Farm-level literature shows that seed could be as much as 50 percent or more of gross production costs (RTIMP 2009; Coyne, Claudius-Cole, and Kikuno 2010). Production projection estimates suggest that gross production will increase from five million tons in the year 2010 to eight million tons in 2025; the net production, after discounting for seed, will be about six million tons.

As pointed out, consumption is projected on the bases of a population growth rate of 1.86 percent per year, a GDP per capita growth rate of 5 percent per year, and a 0.05 income elasticity of demand for yam (FAOSTAT). The projection estimates for Ghana show that aggregate yam consumption will rise from three million tons in 2010 to four million tons in 2025.

The production and consumption projections for Ghana indicate that there will be a surplus of production over consumption of two million tons, which will be available for export. Ghana is a yam export conscious country at the present; ECOWAS policies encourage yam exportation, farmers want to produce for exportation, and exporters are ever ready to take yam out of the hands of the farmers.

The past FAOSTAT data on yam exportation in Ghana are unlikely to account for the entirety of Ghana's yam exportation to West African countries such as Burkina Faso because the regional trade is mostly through informal channels. Earlier in this chapter it was explained that at the Sankare Yaare market in Ouagadougou, Burkina Faso, large quantities of yams were on display for sale. Upon interview,

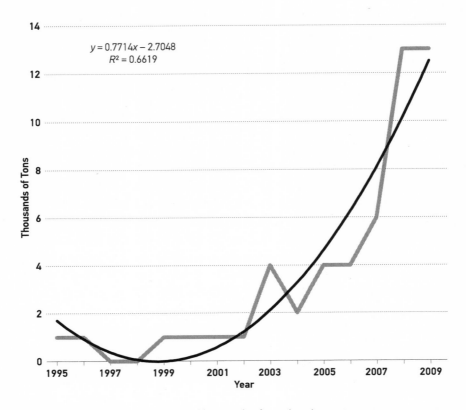

$$y = 0.7714x - 2.7048$$
$$R^2 = 0.6619$$

FIGURE 15. Burkina Faso: Yam import (thousands of tons/year), 1995 to 2009. SOURCE: FAOSTAT.

the sellers reported that all the yams were imported from Ghana and that yams from anywhere else faced low consumer acceptance in the country. It was also explained that at Soukouni Coura de Medine in central Bamako, large quantities of yams were similarly on display for sale. In contrast to the Sankare Yaare market in Ouagadougou, the yams displayed in Soukouni Coura de Medine were mostly imported from Côte d'Ivoire.

This situation means that import demand in Burkina Faso is an important component of Ghana's yam export market. In Burkina Faso from 1995 to 2009, yam importation maintained a strong positive trend, with only minor fluctuations (figure 15).[3] This period of positive trends in Burkina Faso yam importation corresponds to the period of strong positive trends in yam exportation from Ghana, further lending support to the theory that Ghana controls a large share of Burkina Faso's yam importation market.

What is the size of Burkina Faso's yam import market? Projection estimates show that aggregate yam consumption in Burkina Faso will soar from forty thousand tons in 2010 to seventy thousand tons in 2025. High population growth will account for this staggering expansion in yam consumption. A population growth rate of 3.1 percent per year in Burkina Faso is higher than 2.61 in Mali, 1.97 in Nigeria, and 1.86 percent in Ghana (FAOSTAT). The production projection graph is flat at about sixty thousand tons from 2010 to 2025 following the stagnant past production trend. The net production of seed will be even less, at fifty-four thousand tons. In 2025 Burkina Faso will generate a deficit of about sixteen thousand tons, which should be filled from importation sources that will most likely be from Ghana.

Using FAOSTAT, Nathalie Me-Nsope and John Staatz reported yam production has increased, even on a per capita basis, in places like Ghana, Benin, and Nigeria and even held its own in Côte d'Ivoire. They conclude that, while much of the public debate in West Africa is on the increasing dependence on rice and wheat, there seems to be not only a cassava revolution going on but also, for some countries, big increases in apparent consumption of yams (Me-Nsope and Staatz 2013).

Summary

Volumes and values of yam trade in West Africa are unknown because of the lack of recorded data. Global trade in West African yam is driven by growing African populations in diaspora. Regional trade between major and minor producing countries within West Africa is driven by rising populations and incomes in importing countries and by regional ECOWAS policies of development of interstate highways and liberalization of immigration procedures. Standard intervention measures in trade, such as tax breaks and export/import subsidies, are administratively difficult to implement by the poorly performing governments of Nigeria, Ghana, Burkina Faso, and Mali.

Production and consumption projections suggest that West African major yam-producing countries, such as Ghana, will generate significant surplus amounts of yam in the next fifteen years. Minor producing countries such as Burkina Faso will have production deficit over consumption needs. Large surpluses are unlikely to occur in any West African countries other than Nigeria and Ghana, given the low levels of aggregate production in all other countries. In the case of consumption, several other countries in the region like Burkina Faso will generate deficits and will be able to fill such deficits with imports from Nigeria and Ghana. These other countries are unlikely to exhaust surpluses generated in the two major producing countries. The major producing countries will therefore need to be aggressive in sourcing export markets outside the region.

The projections call for investment in measures to expand yam consumption at home, to expand export opportunities, or to do both. Speeding up the rate of improvement of the West African interstate highways will help distribute surplus yams in Nigeria and Ghana to marginal producing countries through export-import trade.

CHAPTER 12

Fuels of Yam Consumption in West Africa

I n frequency terms, Nigeria is the largest yam consumer among the four yam consumption study countries (figure 16).[1] Relatively low frequency of yam consumption in Ghana, the second largest global yam producer after Nigeria, is surprising. Also surprising is the relatively high frequency of yam consumption in Mali, a minor yam-producing country; Mali imports yam mostly from Côte d'Ivoire.

In three of the four yam consumption study countries, the exception being Burkina Faso, frequencies of yam consumption are about even among producing and nonproducing regions. Centre-Ouest, Sud-Ouest, Hauts-Bassins, Cascades, Est, and Centre-Sud are the yam-producing regions in Burkina Faso; the frequencies of yam consumption are higher in those regions than in other regions because non–yam-producing regions of Burkina Faso import yam from Ghana mostly over bad roads.

The point of the observations is that, depending on the condition of market access infrastructure, average frequency of yam consumption is not necessarily higher in producing areas than in nonproducing areas in the yam consumption study countries. This suggests that yam producing households do not necessarily consume yam more frequently than nonproducing urban households, depending on transportation cost. Why? Yam is a high value crop, and low income producers sell

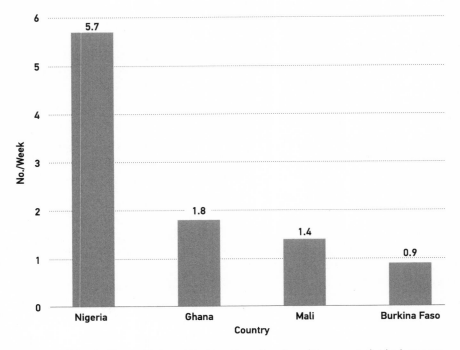

FIGURE 16. Nigeria, Ghana, Mali, and Burkina Faso: Number of times an individual ate yam in a week. December 2012. SOURCE: NWEKE, AIDOO, AND OKOYE. 2013.

their yam and use the money to buy cheaper food crops, such as cassava in Nigeria or maize in Ghana. But movement of yam at high costs can cause consumption frequency to be higher in producing than in nonproducing regions. In Burkina Faso, yams sold in the Sankare Yaare market in Ouagadougou were almost entirely imported from Ghana. Yam consumption study survey travels by bus from Bobo, Burkina Faso, to Kumasi, Ghana, were on a poorly maintained road.

Yam importation from Ghana to Burkina Faso is at a high transportation cost. This can bring about price differentials and therefore differences in yam consumption frequencies between yam-producing and non–yam-producing regions within the country. The Burkina Faso situation illustrates the importance of improved market access infrastructure in yam consumption. Improved transportation facilities will allow people in nonproducing regions equal access to yam as producing areas because yam is produced mostly for sale.

Gender Differences in Yam Consumption

In all four yam consumption study countries, men and women were found to eat yam at about equal frequencies. But men and women ate different yam products at different frequencies where yam food products were diversified, such as in Nigeria and Mali. For example, in Nigeria, men ate pounded yam more frequently than women, and women ate *poundo* yam more frequently than men. In Mali, men were found to eat roasted (grilled) yam more frequently than women, and women ate other yam, such as fried yam (yam cooked in oil), more frequently than men. Therefore, although men and women might prefer different yam food products, gender did not influence the frequency of yam consumption per se.

Yam Food Preparation Technologies

The significance of diverse yam food products for yam consumption frequency is dramatized in the comparison of yam consumption patterns between Ghana and Mali. A marginal producer of yam, Mali shows an average frequency of consumption close to Ghana's, the second largest producer globally. But while yam is eaten almost exclusively in one form, that is, boiled yam, in Ghana, in Mali yam consumption is diversified into boiled, grilled, and fried yams. In Mali, yam, because it is eaten in diversified forms, competes effectively with alternative staples such as sorghum, millet, beans, sweet potato, and potato. The average frequencies of consumption per person per week in Mali are yam, 1.7; sorghum, 1.1; millet, 1.9; beans, 1.3; sweet potato, 1.1; and potato, 1.7.

Physiological Properties of Yam

Comparing the frequencies of consumption of alternative yam food products with their specific substitutes in Nigeria highlights the impact of physiological properties on the frequency of yam consumption. In Nigeria, pounded yam, plantains, and wheat are a different set of substitutes because cooking each of them includes extra preparation cost. Substitution is among pounded yam, pounded unripened plantains, and wheat flour paste because they are eaten the same way, namely as *foofoo*.

Among yam consumers in Nigeria, pounded yam is the ultimate status food, but wheat flour paste and pounded plantain have combined to reduce its consumption frequency to an average of a little more than once a week. The average frequencies are pounded yam, 1.2; wheat, 1.2; and plantain, 3.0 times per person per week. Consumption of pounded yam has dual drawbacks. One drawback is that diabetic patients who are often people with means to engage in the luxury of pounded yam consumption avoid it because they believe it to be high in carbohydrates. These people's preferences for pounded plantain and wheat flour paste are because those products are considered by the consumers to be low in carbohydrates. This reduces the consumption frequency of pounded yam.

The second drawback of the consumption of pounded yam is that yam is unstable in cooked forms such as pounded, boiled, and roasted (grilled) yam or in preprocessed forms such as *poundo* yam. These products lose quality when not warm. Therefore yam is cooked in small quantities in order to be consumed within the shortest period of time to avoid waste, with implications for yam as restaurant food. To eat pounded yam in popular restaurants that serve it, one must go early in the afternoon, or else the prepared pounded yam runs out. In hotel restaurants, pounded yam is served à la carte.

Quality instability is an encumbrance on *poundo* yam consumption. The yam consumption study disclosed that packaged *poundo* yam was not common in Nigerian markets; traders who were interviewed by the yam study investigators explained that demand was low because of high cost and because consumers have access to the real thing, namely pounded yam. The traders further revealed that when sales were delayed, the product deteriorated because *poundo* yam was more susceptible to weevils and fungi than grain flours.

Influence of High Relative Prices on Yam Consumption Frequencies

Different forms of yam are considered different products from a consumer point of view. A prepared form of yam can have a different set of substitutes than another form. In this section, yam consumption is compared with other staples that are yam substitutes as a product and, where applicable, alternative yam products, namely the *poundo* yam, pounded yam, etc., are compared with their specific substitutes.

Yam, as well as several of its staple substitutes, is a seasonal crop in the West African yam belt. Yam consumption data collected following a course route

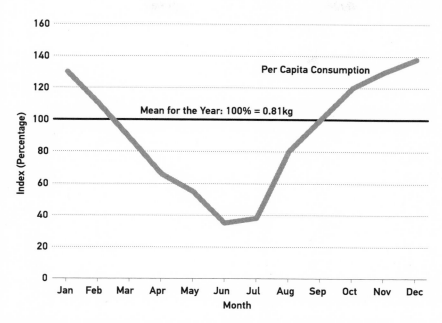

FIGURE 17. Eastern Nigeria: Index of per capita daily yam consumption by month. SOURCE: UGWU 1990.

approach over one calendar year revealed that in eastern Nigeria yam consumption was above average from September of one year to February of the following year and that December was the peak period (figure 17). Similarly, yam price data, also collected using the same course route approach, revealed that yam prices were well below average between October of one year and April of the following year. This data corresponded to harvesting season for most staple crops, except cassava (see figure 12). The yam consumption study survey was conducted right in the middle of the harvesting season, from November 2012 to February 2013, when yam consumption levels were highest and price was lowest.

If a substitute for yam is consumed more frequently than yam during the consumption survey period, the substitute is likely to be consumed more frequently than yam at all other times. Similarly, if a substitute for yam is lower in price than yam during the survey period, the substitute is likely to be lower in price at all other times (figure 12). Cassava is harvested all year round, but its price also declines during other crops' harvesting season in consideration of the other crops' prices;

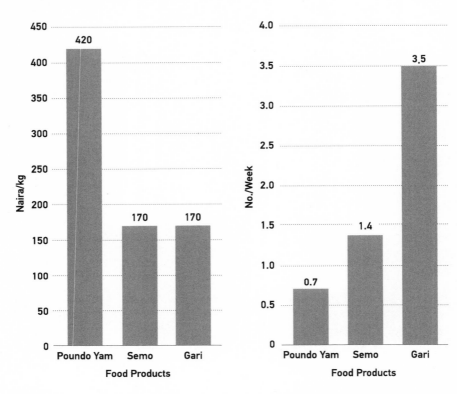

FIGURE 18. Nigeria: Retail prices (₦/kg, ₦160 = US$1) of *poundo* yam, *semo*, and *gari* in Bodija market, Ibadan, December 2012. SOURCE: NWEKE, AIDOO, AND OKOYE. 2013.

FIGURE 19. Nigeria: Number of times an individual ate *poundo* yam, *semo*, and *gari* in a week, December 2012. SOURCE: NWEKE, AIDOO, AND OKOYE. 2013.

that is, the price of cassava declines because of its depressed demand following increased availability and reduced prices of substitutes.

In Nigeria cassava, rice, and plantain are basic substitutes for yam as a staple.[2] In that country, on average, yam is eaten six times a week and cassava seven times. Rice and plantain are eaten less frequently: rice five times a week and plantain three times a week. In Nigeria, yam is prepared as pounded yam, *poundo* yam (industrially prepared yam flour), *amala* (traditionally prepared yam flour), boiled yam, and in other ways. Yam is eaten most frequently as boiled yam, followed by *amala*, pounded yam, and *poundo* yam. *Poundo* yam, *gari* (granulated cassava food

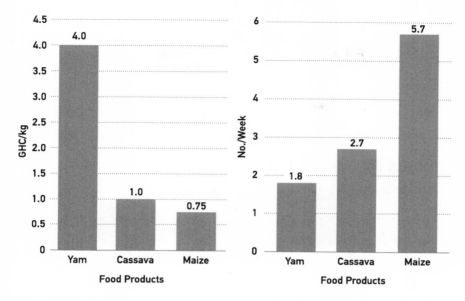

FIGURE 20. Ghana: Retail prices (GHS/ kg, GHS2 = US$1) of yam, cassava, and maize, December 2012. SOURCE: NWEKE, AIDOO, AND OKOYE. 2013.

FIGURE 21. Ghana: Number of times an individual ate yam, cassava, and maize in a week, December 2012. SOURCE: NWEKE, AIDOO, AND OKOYE. 2013.

product), and *semo* (industrially prepared grain flour) are preprocessed products. They are eaten in the same way, that is, as *foofoo*, and are therefore substitutes for one another. Of the three, *poundo* yam is the least frequently consumed; its price at the retail level is over twice the prices of its two competitors (figures 18 and 19).[3]

In Ghana, food staple price information is illuminating; why should Ghana, which produces so much yam, consume yam so infrequently compared with other producing countries such as Nigeria? The retail price information reveals that prices are the main determinant of consumption frequencies of food staples in Ghana (figures 20 and 21). In February 2013 in Kumasi, the retail price of yam was extremely high; four times that of cassava and more than five times the price of maize.[4] These retail price levels adequately explain the following frequencies of consumption of food staples observed in Ghana: yam, 1.8 times per person per week; cassava, 2.7; and maize, 5.7. The inverse relationship between the consumption frequencies of yam and its substitutes and their relative prices is dramatic.

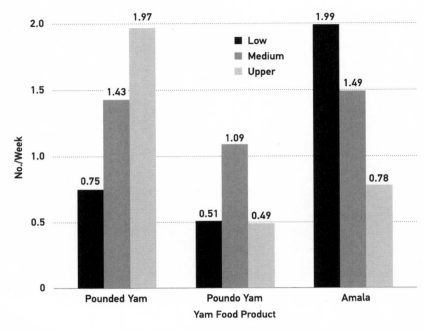

FIGURE 22. Nigeria: Number of times an individual ate yam in a week by relative income group, December 2012. SOURCE: NWEKE, AIDOO, AND OKOYE. 2013.

It has been explained that in Ghana the consumer price of maize is artificially held at a low level by government food crop development programs. The government food crop development programs produce the effect of holding private production costs of affected food crops such as rice and particularly maize at low levels. In the government food development programs yam is sidelined by a lack of improved technologies on which the programs are based.

Income Effect

In West Africa, the yam consumer archetype is best characterized by income and country. A Nigerian in the upper income class defines the yam consumer archetype; a frequent yam consumer picked randomly among West Africans has a high probability of being a Nigerian in the upper income class. In all the four yam consumption study countries, the higher the income group of the consumer, the higher the frequency of his or her yam consumption. But the Nigerian experience reveals that income groups discriminate with regard to which yam product they consume; in that country the consumption frequency of pounded yam increases while the consumption frequency of *amala* decreases from the lower through middle to upper income groups (figure 22). *Amala* appeals to the middle income group because its price (₦210/kg) is half the price of *poundo* yam (₦420/kg) but is still higher than the prices of close substitutes, namely *semo* (₦170/kg) and *gari* (also ₦170/kg).

This powerful association between yam consumption frequencies and income groups lends credence to the earlier observation that yam consumption is driven by market forces.[5] An increase in consumer income will increase yam consumption, particularly in nonproducing areas including importing countries.

Summary

The yam consumer can be characterized best by their country and income group; a randomly picked West African who consumes yam frequently is likely to be a Nigerian in the upper income group. Yam consumption is highly sensitive to consumer income: the higher the consumer income, the higher the frequency of yam consumption. It is also sensitive to relative prices: the lower the price of yam relative to substitutes, the higher the frequency of yam consumption.

The implication of these associations for the future of yam is unambiguous: the future of yam is definitely bright. It will be even more so if R&D measures are put in place to make the crop price competitive with alternative food staples. The positive relationship between consumer income and yam consumption frequency means that yam has a high market demand. The equally strong inverse relationship between price of yam relative to its substitutes means that technologies that drive down yam production cost and reduce the price of yam to consumers will surge yam consumption frequency. Therefore, expanded production achieved at a reduced cost will increase farmer income through increased sales and increase consumer income through reduced prices. Improved yam food preparation technology will further promote yam consumption frequencies in the region.

CHAPTER 13

Yam in the Igbo Cultural Rite of Marriage

I n the literature, New Yam festivals dominate the discussion of cultural rites in which yam plays a pivotal role in producing areas. D. G. Coursey's review of literature on the subject of New Yam festivals suggests that detailed studies of the festivals, which are widely celebrated, is scanty and old (Coursey 1967). Introductions in several, more recent, West African yam literatures acknowledge that yam has an important role in the culture of the people in major producing areas, obviously in reference to Coursey. After the introduction, most of such literatures have little, if anything, to say on the subject (Hahn et al. 1987). Although the role of yam in culture is mentioned, it is not discussed.

Coursey referenced the handful of early detailed studies as those of R. S. Rattray among the Ashanti people of Ghana and those of C. K. Meek and D. Forde among the Igbo and Yako peoples of eastern Nigeria (Rattray 1923; Meek 1937; Forde 1964). Coursey also referenced other scholars who mentioned the festival but did not elaborate on the manner in which it was performed, such as L. Tauxier, J. Miege, and J. L. Boutillier. These authors referred to the performance of New Yam festivals in parts of Côte d'Ivoire, while A. B. Ellis wrote about Togo and N. W. Thomas, Benin Republic (Tauxier 1932; Miege 1954; Boutillier 1960; Ellis 1890; Thomas 1910). In Nigeria, N. W. Ellis and R. E. Dennett made references to the performance of New

Yam festivals among the Yoruba people, R. C. Abraham among the Tiv people, and P. A. Talbot among the Kalabari people (Ellis 1894; Dennett 1910; Abraham 1940; Talbot 1932). Coursey noted that although some of these literature evidences were found several years earlier, New Yam festivals were still held in most of the places up to the time of the publication of his book in 1967 (Coursey 1967).

Significance of the Use of Yam in Cultural Rites

Among the Igbo people of Nigeria, the significance of the New Yam festival is as a thanksgiving festival similar to the popular American Thanksgiving. Turkey is the central ritual object in the American Thanksgiving because it was common in the wild and pioneering European immigrants survived on it. Worldwide, different cultures celebrate nature's bounties in different ways and with different materials that are commonly available or, for one reason or another, were important in the history of the people. Igbo land is considered the white yam's center of origin, which means that the white yam evolved there and sustained the first people in the area.

The New Yam festival can hardly be described as a passing event among Igbo people. Every year around the months of August and September, Igbo people residing in different parts of Nigeria return to their villages in large numbers to participate in New Yam festivals. In some countries outside Nigeria, Igbo people in diaspora celebrate the New Yam festival with characteristic loudness in major cities such as Johannesburg, Chicago, London, and Paris.

Apart from the New Yam festival with its thanksgiving significance, yam is a central object in several other cultural rites of thanksgiving and rites of passage, petition, and appeasement with deep significance in several producing areas across the world. For example, in "Abelam: Giant Yams and Cycles of Sex, Warfare and Ritual," Richard Scaglion, an American anthropologist, reported that for the Abelam people—a New Guinea tribe—warfare and other social activities, as well as some private activities, included the use of giant yams. These giant yams can be as large as ten feet or longer, around which members of the tribe organize their lives (Scaglion 2007).

During the yam growing season the Abelam would not fight because if they did, yam, which abhorred violence, would not grow big tubers. After yams were harvested, yam festivals were organized in different villages, and enemies cooperated by suspending hostility. Enemies would also visit one another's villages to feast together, to measure and inspect one another's yams, and to exchange yam. Individuals gave the largest and best of their yams to their archrivals who strived to produce a bigger yam to exchange with the same rivals the following season. Yam exchanges were alternatives to physical violence as means of dealing with social conflicts; through yam exchanges, conflicts were redirected into a socially acceptable, nonviolent channel.

Ponapeans of Ponape in the Pacific Islands considered yam to be their most important crop although the crop did not contribute to their food security as much as did bread fruit (Mahony and Lawrence 1959). Yam was not bought or sold, and production techniques were designed to produce large tubers rather than the highest yield per unit area or per man hour of labor. The larger the tuber size a man grew, the faster he could move up in the title system of the people (Mahony and Lawrence 1959).

An Igbo man who aspired to attain a position of respect in his community had to take one or more chieftaincy titles, which became progressively harder and more expensive as he proceeded higher on the social ladder (Coursey 1967; Achebe 1988, 2012). These titles were concerned with the degree of success in yam production in terms of how much in total quantity and in size of individual tubers the man was able to show. This social practice had important significance in community governance; in a village setting, a successful yam farmer was a community mobilizer to whom community members listened. Nowadays such people are used by extension agents as resource persons for delivery of new farm technologies to farmers.

Contribution of this Chapter to Use of Yam in Cultural Rites

The use of yam in cultural rites other than New Yam festivals is common among major producing centers in West Africa. For example, the yam consumption study revealed that Dagomba people who live near Tamale in northern Ghana celebrate yam during festivals, chieftaincy titles (enskinning), and sacrifices to the gods, with marginal differences from the Igbo people. The analysis of marriage rites among the Igbo people will sufficiently demonstrate the impact of cultural rites on yam production and consumption in West Africa. Focusing on only the marriage rites of the Igbo people will circumvent repetitions that could be uninteresting to readers since procedures are similar across different rites and across yam-producing centers in West Africa.

Igbo Rite of Marriage

In Igbo practice, there are two parts to the rite of marriage, namely ritual and celebration; celebration is similar to a party, which takes place after the ritual. At the party, drinks and food, including yam, are served. This means that celebrations during the performance of marriage rites contribute to yam consumption. Of more interest in the discussion of those rites as important factors in promoting yam production and consumption is yam as a ritual object.

The marriage ritual procedure demonstrates the supremacy of communal interests over private interests. In the Igbo village of Ukpo, Dunukofia, the rituals call for the largest yam tubers available in the market. At marriage, a specified number of such yam tubers are supplied by the family of the bridegroom. After the leadership of the bride's extended family confirms the adequacy of the yams in terms of number and size of tubers, the yams are distributed in specified numbers to the oldest members of the bride's extended family. The balance of the yams is cooked, along with accompanying animal sacrifice, such as a goat, for all members of the extended families to share. The procedure demonstrates the upbringing of children as a communal responsibility; the extended family that helped to raise a girl shares in her marriage ritual.

Marriage does not take place without this ritual; the bridegroom must be accompanied by representatives of his extended family, and representatives of the bride's extended family must be present to receive them. Both sides must endorse the marriage for it to take place, and this is for good reason. If either the bride's or the groom's family is not properly represented, it is understood that a family member is defiant. This is how the various rites constitute the glue that binds a community together. For the marriage to proceed, family on each side must settle differences with their community as a whole.

Often novices, that is, people who do not sufficiently understand the role of yams in the Igbo marriage rites, draw a parallel between yam tuber size and masculinity; that could have been part of the significance of yam in the Igbo marriage rites especially in the past when food crops were not widely traded in the villages and people used the yam they produced for the marriage ceremonies. Among herdsmen brides are priced in number of cattle; a highly prized bride calls for a large number of cattle at marriage.

Size of Ceremonial Yam Market

How large is the ceremonial yam market? Is demand for yam in the various rites of passage, thanksgiving, petition, and appeasement large enough to make a significant impact on yam consumption? The farmer groups interviewed for the yam consumption study pointed out the unusualness of people looking for the largest yam tubers in the market to cook at home because it is wasteful. In a matter of hours, the cut surface of a tuber begins to oxidize turning to unusable cake that must be peeled off the next time the yam is used. In a few days rot begins to form from the oxidized surface and spreads to the rest of the tuber. Therefore when people shop for yam to cook at home, they purchase a tuber the family can eat at one meal to avoid losses.

The farmer groups also pointed out the usualness of yam producers to aim at producing the largest tubers they can. The farmers' desire to grow the largest possible tubers is driven by the ceremonial yam market demand, which is set against the home consumer preference for medium-sized tubers. This is incontrovertible evidence that the ceremonial yam market is in fact large.

To press further their argument that the amounts of yams used for cultural rites each year were sufficient to constitute a significant demand, the farmer groups pointed out that each year the numbers of marriages, births, and funerals in both rural and urban settings were high. These are in addition to a litany of heathen shrines that demand and receive tributes of yam daily for thanksgiving, petition, and appeasement. The farmer groups pointed out that rites of passage, thanksgiving, petition, and appeasement are also performed in Christian churches. They point at the number of yams presented in each church every Sunday for purposes of petition and penitence and for thanksgiving following marriages, births, and funerals as well as numerous other events.

Sustainability of the Ceremonial Yam Market

The size of the ceremonial yam market is convincingly large enough to constitute an effective demand for yam. But what about the sustainability of such a market over time; in other words, are the cultural practices sustainable in the future? In his latest classic, Wole Soyinka, a Nobel Prize–winning Nigerian novelist and playwright, laments the erosion of African culture by both Christian and Islamic civilizations

(Soyinka 2012). But, as noted earlier, rites of passage, thanksgiving, petition, and appeasement are practiced in Christian churches with yam in producing areas. In addition, as also pointed out earlier, the ceremonial yam market is sustainable because the cultural rites in which ceremonial yams are used are sustainable as the rites are important in enforcing compliance to communal interests and in helping support community governance.

What does a medal of honor have to do with ceremonial-sized yam production among the Igbo people?[1] Interest in the production of ceremonial-sized yams is institutionalized by practice. In the past, Igbo yam-producing communities had a *di ji* (master yam farmer) or *eze ji* (king among yam farmers), who was a community leader. His voice was heard in his community. Elevation to that status was derived from the total quantity and size of his individual tubers. Today in this and other major yam-producing areas, trade fairs emphasize tuber size; the trophy goes to the largest tuber on display.

Production of Ceremonial Yam and Technology Development

The discussion of whether there is an association between the farmers' quest for the largest possible tuber size and yam technology development aims to provoke thinking on the topic. In this section, questions are posed without an attempt at postulating any answers at all because of the lack of empirical information on the subject.

Is the emphasis on producing large tubers an impediment to technology change in yam production? Farmer groups interviewed in the yam consumption study were adamant that there has been no change in yam production technology in their living memory. Given their environmental conditions, including soil structure and fertility, the farmers adopt cultivars and agronomic practices that produce the largest-sized tubers possible. Is there a conflict between yam technology development and an emphasis on producing the largest possible tuber size?

Farmers aim to grow the largest tubers they can produce because of the large ceremonial yam market. Like Ponapeans, yam farmers in West Africa define yield in terms of tuber size and adopt cultivars and agronomic practices likely to produce large tubers, overlooking scientific evidence that aggregate yield per unit area or per man hour of work is higher in producing tubers that are smaller than the ceremonial yam tubers.

Yam cultivars are unequal in their tolerance to diseases and pests. Cultivars that are tolerant to yam pests and diseases may not yield large-sized tubers. For example, a cultivar called *abii* in the Otuocha area of eastern Nigeria is known to be, to some extent, tolerant to certain nematodes of yam but yields smaller-sized tubers than other white yam cultivars. The *abii*'s small-sized tuber attribute may hinder its use in breeding even though it has a nematode tolerant attribute. Table 7 shows that *abii* is at the bottom of the list of the top five most popular cultivars, listed in order of popularity, grown in the Otuocha area.

In spite of empirical evidence that planting yam in ridges produces a higher aggregate yield per unit area, is labor saving, and may be easier to mechanize, farmers everywhere in the region continue planting yam in mounds. Planting in yam ridges would permit higher plant population and facilitate implementation of agronomic practices. Is the need to produce large-sized tubers responsible for farmers' dogmatic adherence to planting yams in mounds? Is there a relationship between yam seedbed type and the size of the tuber produced?

The yam consumption study reveals that consumers pay premium price for large tubers (see figure 13). Ceremonial yam production entails high costs of large seed yam, large mounds, large stakes, and lower aggregate yield per unit area. Whether the difference in price compensates for the additional costs of ceremonial yam production is an empirical question that begs for investigation.

In yam technology development, trade matters because market demand drives the technology development for a product. Kintamkpo (Ghana) area farmers narrowed the yam varieties they produce to what the market demands. Growing demand for yam outside of West Africa provides hope for an increase in demand for nonceremonial yams. This is because yam importers outside of West Africa may not have ceremonial value for large tubers. Would higher demand from such consumers discourage farmers in West Africa from dogmatic adherence to production methods adapted to generating the highest possible tuber-sized yams?

Summary

This chapter describes the use of yam as a central object in the marriage ritual among the Igbo people of Nigeria, using the practice in the town of Ukpo, Dunukofia, in eastern Nigeria as an example. Large ceremonial yams that are to the satisfaction of the bride's family must be provided by the bridegroom's family. More importantly, members of the extended family of the bridegroom must accompany him to meet the members of the bride's extended family in order to be received. The extended families will not participate in this ceremony if the bride or the bridegroom or either of their respective families is not at peace with their respective communities; any disagreement must be settled ahead of marriage. This way, yam during marriages helps maintain communal rules, peace, and order in Igbo communities.

The demand for large tubers for ceremonial purposes is considerable because there are many ceremonies in Igbo villages and urban settings. The ceremonies are also sustainable because of their value in maintaining peace and order in the communities and because some of the ceremonies are engaged in by young people and in both Christian churches and heathen shrines.

The influence of the farmers' quest for large ceremonial yams on yam technology development is uncertain. Ceremonial yams are expensive to produce because they need large seed yams, large mounds, and tall stakes. Consumers pay premium price for large tubers to use during ceremonies.

CHAPTER 14

Synthesis

In Nigeria and Ghana yam is mostly produced in villages that are remote from urban areas and have poor road access to market centers. The low quality of life in the yam-producing villages, especially in Ghana, is on graphic display in the dependence of the village people on puddles and rivulets for drinking, cooking, and washing water supplies. The puddles and rivulets are stagnant in dry seasons and could be a vehicle for waterborne diseases. Zero or little formal education predisposes the mainline farmers in both Nigeria and Ghana to unorthodox yam production practices, such as reliance on rituals in preference to trading in a critical purchased input, namely seed yam. The farmers with little or no formal education blame ritual practices by neighbors for the failures of their yam crop that may actually be caused by yam plant pests and diseases, low soil fertility, or bad weather. The farmers seek solution to the aforementioned problems in ritual practices.

Among the most critical constraints in yam production in West Africa is the practice of shifting cultivation; this occurs mostly in Ghana. The practice exposes the farmers in that country to unproductive and tortuous commuting mostly on foot over long distances between their homes and the yam fields. The practice of shifting cultivation, which is rooted in the farmers' continuous search for fertile land, low yam pest and disease pressure, and stake trees, has negative implication of environmental degradation.

The yam marketing system works against the farmers, forcing them to live in penury in spite of their backbreaking yam production engagement creating the situation whereby consumers pay high prices yet producers live in poverty. Determination of the obstacles against farmers receiving compensation that is commensurate to their efforts and identification and implementation of policy measures to effectively eliminate those obstacles are essential in order to bring down the high level of poverty in yam-producing communities in West Africa.

High Labor Cost in Yam Production

Certain agronomic practices that are peculiar to yam make yam production labor-intensive in the extreme. On a per unit area basis aggregate yam production labor is one and a half times that of cassava. This is because yam is grown in mounds and planted after fallow more frequently than cassava; the nuisance of weeds is higher in yam production than in the production of cassava; and yam is staked and mulched, while cassava is not. Yam researchers are challenged to fathom the underlying reason for the farmers' dogmatic reliance on mounds for yam production even though available scientific evidence suggests planting in ridges is laborsaving, easier to mechanize, and gives higher aggregate yield per unit area or per unit of labor input.

Yam researchers have shown that staking yam produced a yield advantage over not staking yam and that PVC pipes and plastic mesh proved to be suitable alternatives to wood, yet farmers dogmatically stick to using wooden stakes. As yam production expands, demand for wood for stakes increases, exerting pressure on the environment. Some entrepreneur will one day see a business opportunity in yam stake farming and supply the growing demand. There is already a growing market for yam stakes in parts of Nigeria, but supplies are collections from the wild.

There is a high potential for cost reduction in the use of mechanical laborsaving technologies in yam production. Reliance on imported prototype machines has a low chance of success because yam is a West African crop that is hardly understood by engineers from outside the region. Designing laborsaving mechanical technology for yam production and harvesting calls for imaginative thinking because of yam's peculiar characteristic agronomic practices such as mound making, staking, and harvesting by digging vertically. Such a design has need for a tripartite effort among engineers, scientists, and farmers who understand the yam culture.

Seed Yam Technology

The high cost of seed yam is a major bottleneck in yam production because the edible tuber—the crop—is used as seed. The most advanced of the new seed yam technologies, the minisett technique, has a low adoption rate. Farmers have continued to practice old technologies that result in low multiplication rate for seed yam as well as cause a depressed tuber yield because the methods recycle seed yam, increasing the amount of seed yamborne pests and diseases.

Without question, considerable progress has been made in seed yam technology development, but numerous challenges are still outstanding. Some promising novel technologies, which are potential alternatives to tuber as seed yam, are at laboratory and greenhouse stages of development. Aeroponics-system and tissue-culture technologies, which rely on stem cutting, have been proved to be technically feasible; in fact the aeroponics-system technology qualifies as a breakthrough. But effectiveness of any seed yam multiplication technique, no matter how rapid the multiplication rate may be, depends on the control of seed yam–borne pests and diseases, especially the nematode. Yam does not multiply rapidly because of the hindrances of pests and diseases.

Significant strides in seed yam technology development have occurred with minimal participation of the private sector other than the farmers themselves. Advocacy among political leaders and policy makers would help to create a conducive macroeconomic environment to provide incentive for private investors to take advantage of the new technologies to produce and supply farmers with quality-declared seed yams at affordable prices.

The Challenge of Nematodes in the Yam Food Sector

Nematodes are a dread to stakeholders at all links in the yam value chain, including seed yam multiplication and yam production, storage, marketing, and consumption.

The nematode pests confine the West African yam belt to less than what could be potentially productive areas. Extensive production methods that farmers adopt as control measures are inefficient in their use of land resources, and they constitute a serious threat to the environment in terms of deforestation. The dread of the nematode pest reduces the yam marketing channel by eliminating rural assembly markets to the farmers' disadvantage. The problems do not permit bulk buying of yam by housewives, putting yam consumption at a disadvantage and favoring grains and processed cassava products instead. The nematode pest problem becomes an impediment to the diffusion of new technologies in the yam food sector, and yam becomes sidelined in national food crop development programs.

Measurement of the damage done to yam by nematodes in terms of yield and storage loss constitutes gross underestimation. Damage caused by the pest problem to yam in particular and to the ecology in general should be defined broadly. This definition should include the contributions of the problem to the high cost of seed yam, continued practice of yam production under extensive methods, and narrowing of yam production areas to limited ecological zones. The damage should also be defined to include negative impacts on yam marketing and consumption, as well as the sidelining of the crop in national food policy programs.

In summary, success in arresting the problem of nematodes in the yam food sector will enable

1. the take-off of the diffusion of various yam technologies, including novel seed yam technologies and hybrid yam varieties;
2. expansion of yam production in marginal humid ecologies;
3. yam cultivation by intensive methods;
4. improved yam marketing efficiency, empowering producers at the price negotiating table with yam traders;
5. bulk purchase of yam by consumers as they are already able to do with grains and processed cassava products; and
6. bringing yam into central focus to benefit from national food policy programs.

Presently, cultivated yam varieties are susceptible to nematodes. Research-recommended control measures for the pests and diseases are palliative and therefore not effective. Effective solutions could require radical programs such as biological control and transgenic engineering.

Yam Crop Improvement Research

Formal yam crop improvement research in West Africa started when the IITA's yam research program was established in 1971. At the time, newly established yam research programs in West Africa had only morphology characterization information provided by botanists as the genetic resource to start with for yam crop improvement research. But within three years of the establishment of the programs, the researchers achieved breakthroughs. These breakthroughs included inducing flowering in major cultivated yam varieties and synchronizing the flowering period in male and female yam plants, thereby setting the stage for an ambitious yam breeding program. Thirty years after the establishment of the research programs in the early 1970s, researchers made the initial releases of new yam varieties to farmers in 2001.

Successes achieved in yam crop improvement research, some within a short period of time, are convincing evidence that yam responds positively to investment in research. It takes a long time—three decades in the case of yam in West Africa—to come up with new varieties, especially when there is no genetic material developed earlier to build on. Researchers often do not have control of their work schedule, as administrators can stop the researchers' work at any time. From 1988 to 1991, IITA halted the yam crop improvement research at a point of high optimism raised among the researchers by their breakthroughs in inducing flowering in major cultivated yam varieties and in synchronizing the flowering period in male and female yam plants. While some breeders are promoted to senior administrative positions in the hierarchy of their institutions, administrators prematurely terminate the contracts of some other breeders.

Diffusion of Hybrid Yams in Nigeria

The diffusion rate of the hybrid yam varieties released to farmers in Nigeria more than ten years ago is low. This is because the new hybrid yams do not address some of the most critical needs of farmers and consumers, and the precondition for their adoption is not in place. The problem of yam pests and diseases is among the most serious of the farmer and consumer felt needs. The solution to that problem is a precondition for the diffusion of most new technologies in the yam food sector. Ability of a new technology to solve a felt need and the preexistence of complementary technologies, if any, are necessary conditions for adoption; promotion serves to speed up the momentum of diffusion.

Most agricultural technologies require precondition for adoption; availabilities of chemical fertilizers and irrigation water were preconditions for adoption of the Green Revolution wheat, rice, and maize in South America and Asia in the 1960s and 1970s. The mechanical cassava grater for preparing *gari* was in place in Nigeria when the IITA's high-yielding, mosaic resistant TMS cassava varieties were developed and released to farmers in the late 1970s. Control of major yam pests and diseases, such as nematodes, is a precondition for the diffusion of high-yielding hybrid yams in Nigeria.

More than promotion, the prospect for diffusion of hybrid yams in Nigeria depends on the establishment of precondition for adoption—namely control of major yam pests and diseases, especially nematodes, which constitute the biggest impediment against the adoption of hybrid yams. The decision to invest in multiplication and distribution to farmers of the hybrid yams at the present stage of yam crop improvement research should be made with caution. The susceptibility of the hybrid yams to pests and diseases could lead to an embarrassing failure that could create situations of reverse adoption, farmer disaffection with research, and loss of funds invested in diffusing the hybrid yams.

Postharvest Matters

Shortage of market information, dread of yam pests and diseases, reactionary pricing strategies, and high transportation costs are principal issues of concern in yam marketing in Nigeria and Ghana. Apart from these, other deficiencies that need fixing include the lack of standards and measures and the practice of displaying yams out in the open, under the sun and in the heat.

The yam marketing channel is weak in terms of rural assembly and wholesale warehousing; in some cases urban wholesale traders bought at the farms and in other cases farmers sold at the urban wholesale markets. Under the latter practice, the farmer who paid transportation costs to get from the farm to a long distant urban wholesale market center for a truckload of yam that is perishable is in a weaker bargaining position for price because the cost to the wholesale trader who arrives at the farm empty handed is lower. Therefore the insufficiency of the rural assembly traders, caused by the fear of pests and diseases in the yam marketing channel, does not favor yam producers. Similarly, the absence of speculative function in the wholesale market works against consumers who pay high prices during lean periods. These lean periods occur because of the marketing system's inability to even out seasonal supplies by speculative warehouse storage, also as a result of the fear of yam pests and diseases. As in yam production, the dread of pests and diseases is a major bottleneck in yam marketing in West Africa.

The odds of using yam as feed and industrial raw material depend on the implementation of measures that will drive down the cost of yam relative to alternative crop sources of the raw material. An increased consumption of yam in processed forms depends on the development and dissemination of technologies for processing yam into products of present consumption forms. Improving the quality of dried yam tuber flour to make it competitive in quality with its unprocessed alternative and reducing the cost of the flour to make it price competitive with substitutes prepared from other commodities are other technologies that need to be developed to increase consumption.

The inability of existing yam storage technologies to hold yam from one harvest to another creates seasonality in supply and market prices, leading to a higher use at harvest than at other times. An additional reason for the high use of yam at harvest is the fear of damage; the rate of damage by pests and diseases is highest for tubers with physical injuries, which are often sustained during the harvesting process.

Fuels of Yam Consumption

There is a direct association between the frequency of yam consumption and the consumer's income group and an inverse association between the frequencies of yam consumption and the retail market price of yam relative to the prices of its substitutes. The implication of these associations for the future of yam is unambiguous; the future of yam is definitely bright, more so if R&D measures are put in place to make the crop's price competitive with alternative food staples. The relationships that are dramatic and consistent across the yam consumption study countries underscore the argument that an increase in consumer income in the representative countries would positively impact yam consumption. Improvement in the road networks within yam-producing and yam-consuming countries would also have a positive impact on the frequencies of yam consumption.

What is the implication of these strong relationships? The positive relationship between consumer income and frequency of yam consumption means that yam has high market demand. The equally strong but inverse relationship between the consumption frequencies of yam and its substitutes and their relative prices means that technologies that will drive down yam production costs and reduce the price of yam to consumers will increase yam consumption. Expanded production achieved at a reduced cost will increase farmer income through increased sales and increase consumer income through reduced prices. Improved yam food preparation technology will further promote yam consumption in the region.

Yam in Ceremonies

Demand for yam for use as a ritual object in cultural rites of passage, thanksgiving, petition, and appeasement practiced in major producing and consuming centers is high enough to have a significant effect on yam consumption; the market is sustainable because the rites are sustainable. The influence that the ceremonial yam market may be exerting on redundant yam production technology calls for investigation. Farmers focus on producing large-sized yam tubers to satisfy the ceremonial yam market. But it is not known if farmers' reluctance to change their production technology has anything to do with their quest to produce the largest yam tubers possible given their environmental conditions.

Participants in Activities that Contributed Information Reported in this Book

Information reported in this book was derived from several sources that included the following activities: YIIFSWA Baseline Survey, Yam Consumption Patterns in West Africa Study, and the YIIFSWA 2012, 2013, and 2014 Annual Progress Review and Planning Meetings. The following lists include the individuals who participated in these activities.

The YIIFSWA Baseline Survey Team

Tahirou Abdoulaye
Adebayo Akinola
Robert Asiedu
Daniel Dompreh
Nobert Maroya
Djana Mignouna

Sakina Moro
Felix Nweke
Adewumi Okulaja
Adenike Sanusi
Issahaq Sulaiman
Idowu Temitope

The Yam Consumption Patterns in West Africa Study Team

Robert Aidoo
Sibiri Biriba
Makan Fofana
John Ikeorgu

Felix Nweke
Benjamin Okoye
San Traore

Participants in the YIIFSWA (Yam Improvement for Income and Food Security in West Africa) 2012, 2013, and 2014 Annual Progress Review and Planning Meetings[1]

Louise Abayomi
Tahirou Abdoulaye
Sunday Abimiku
Oladeinde Adebosola
Temi Adegoroye
Hans Adu-Dapaah
Evelyn Adu-Kwarteng
Samuel Adzivor
Clement Agada
Rhode Ahlonsou
Robert Aidoo
Beatrice Aighewi
Adebayo Akinola
Adeola Akintoye
Sunday Aladele
Oladeji Alamu
Sara Alexander
Denise Amenga
David Annang
Antonio Lopez-Montes
Ernest Asiedu
Robert Asiedu
Sheila Assibey-Yeboah
Robert Asuboah

Henry Asumadu
Owati Ayodeji
Morufat Balogun
Sibiri Birba
Paul Boadu
Emmanuel Bobobee
Nsaidoo Brafu
Emmanuel Chamba
Danny Coyne
Eric Danquah
Alexandre Dansi
Cevas Dormediameo
Zidafamor Ebiarede
Okechukwu Eke-Okoro
Emmanuel Ekundayo
Kenneth Ekwe
Stella Ennin
Hussein Etudaiye
Olurotimi Famodile
Mohammed Ibrahim
John Ikeorgu
Keane Jules
Abdoulaye Kah
Regina Kapinga

Ulrich Kleih

Amatevi Klutse

Raoul Klutse

Hilde Koper

Julie Kragh

Lawrence Krampa

Lava Kumar

Bentil Kwame

Walter Leke

Katherine Lopez

Kristen MacNaughtan

Omodamiro Majekodunmi

Hernan Manson

Norbert Maroya

Nora McNamara

Djana Mignouna

Sakina Moro

Steve Morse

Godfrey Mulongo

Fadel Ndiame

Tuu-Van Nguyen

Chukwuemeka Nkere

Emmanuel Nwachukwu

Felix Nweke

Ndidi Nwuneli

Oiwoja Odihi

Samuel Ogbe

Anthony Olatokun

Folake Olokun

Oguntade Oluwole

Celestina Omohimi

Adewuyi Omomuwi

Francis Onukwe

Ogedengbe Opeyemi

John Orchard

Cecil Osei

Kingsley Osei

Angela Osei-Sarfoh

Emmanuel Otoo

Taiwo Oviasuyi

Oyelami Olayemi

Gboyega Pelemo

Dai Peters

David Phillips

Marian Quain

Albert Quainoo

Klutse Radul

Debbie Rees

Sammy Sackey

Lydia Sasu

Samuel Sey

Adamu Shuaibu

Issahaq Sulaiman

Peter Uvere

Takagi Watanabe

Jamaldinne Zainab

Ebiarede Zidafamor

List of Hybrid Yams Released to Farmers in Nigeria

CROP NAME	S/N	VARIETY NAME	ORIGINAL NAME	NATIONAL CODE	ORIGIN/SOURCE
Yam	418	TDr 89/02677	TDR 89/02677	NGDR-01-1	NRCRI, Umudike/IITA, Ibadan
Yam	419	TDr 89/02565	TDR 89/02565	NGDR-01-2	NRCRI, Umudike/IITA, Ibadan
Yam	420	TDr 89/02461	TDR 89/02461	NGDR-01-3	NRCRI, Umudike/IITA, Ibadan
Yam	421	TDr 89/02665	TDR 89/02665	NGDR-03-4	IITA, Ibadan/ NRCRI, Umudike

DEVELOPING INSTITUTE	BREEDER/ COLLABORATING SCIENTISTS	OUTSTANDING CHARACTERISTICS	YEAR OF RELEASE	YEAR OF REGISTRY
NRCRI, Umudike/IITA, Ibadan	S. K. Hahn, R. Asiedu, & G. C. Orkwor	Stable yield, very good cooking and pounding qualities, cream tuber parenchyma, 25% tuber dry matter content.	2001	2001
NRCRI, Umudike/IITA, Ibadan	S. K. Hahn, R. Asiedu, & G. C. Orkwor	Stable yield, very good cooking and pounding qualities, cream nonoxidizing parenchyma, 35% tuber dry matter.	2001	2001
NRCRI, Umudike/IITA, Ibadan	S. K. Hahn, R. Asiedu, & G. C. Orkwor	Stable yield, very good cooking and pounding qualities, cream parenchyma, 26.7% tuber dry matter.	2001	2001
IITA, Ibadan/ NRCRI, Umudike	S. K. Hahn, R. Asiedu, & G. C. Orkwor	Stable yield, very good cooking and pounding qualities, cream nonoxidizing parenchyma, 35.3% tuber dry matter.	2003	2003

CROP NAME	S/N	VARIETY NAME	ORIGINAL NAME	NATIONAL CODE	ORIGIN/SOURCE
Yam	422	TDr 89/01213	TDR 89/01213	NGDR-03-5	IITA, Ibadan/ NRCRI, Umudike
Yam	423	TDr 89/01438	TDR 89/01438	NGDR-03-6	IITA, Ibadan/ NRCRI, Umudike
Yam	424	TDr 95/01924	TDR 95/01924	NGDR-03-7	IITA, Ibadan/ NRCRI, Umudike
Yam	425	DRN 200/4/2	DRN 200/4/2	NGDR-08-8	NRCRI, Umudike
Yam	426	TDa98/01176	TDa98/01176	NGDA-08-9	IITA, Ibadan
Yam	427	TDa98/01168	TDa98/01168	NGDA-08-10	IITA, Ibadan
Yam	428	TDa98/01166	TDa98/01166	NGDA-08-11	IITA, Ibadan
Yam	429	TDr 95/19158	TDr 95/19158	NGDR-09-12	IITA, Ibadan
Yam	430	TDr 89/02602	TDr 89/02602	NGDR-09-13	IITA, Ibadan
Yam	431	TDr 89/02660	TDr 89/02660	NGDR-09-14	IITA, Ibadan

DEVELOPING INSTITUTE	BREEDER/ COLLABORATING SCIENTISTS	OUTSTANDING CHARACTERISTICS	YEAR OF RELEASE	YEAR OF REGISTRY
IITA, Ibadan/ NRCRI, Umudike	S. K. Hahn, R. Asiedu, & G. C. Orkwor	Stable yield, very good cooking and pounding qualities, white nonoxidizing parenchyma, 29.8% tuber dry matter.	2003	2003
IITA, Ibadan/ NRCRI, Umudike	S. K. Hahn, R. Asiedu, & G. C. Orkwor	Stable yield, very good cooking and pounding qualities, white nonoxidizing parenchyma, 29.3% tuber dry matter.	2003	2003
IITA, Ibadan/ NRCRI, Umudike	S. K. Hahn, R. Asiedu, & G. C. Orkwor	Stable yield, very good cooking and pounding qualities, white nonoxidizing parenchyma, 32.8% tuber dry matter.	2003	2003
NRCRI, Umudike	E. C. Nwachukwu	High yielding, pest and disease tolerant, very good for fufu, frying, and boiling.	2008	2008
NRCRI Umudike	R. Asiedu & C. N. Egesi	High yielding, pest and disease tolerant, good for pounded yam, frying, and boiling, suitable for both rainy- and dry-season yam production.	2008	2008
NRCRI Umudike	R. Asiedu & C. N. Egesi	High yielding, pest and disease tolerant, good for pounded yam, frying, and boiling.	2008	2008
NRCRI Umudike	R. Asiedu & C. N. Egesi	High yielding, pest and disease tolerant, good for pounded yam, frying, and boiling, suitable for both rainy- and dry-season yam production.	2008	2008
NRCRI, Umudike	R. Asiedu	High yielding, pest and disease tolerant, very good for fufu, frying, and boiling.	2009	2009
NRCRI, Umudike	R. Asiedu, J. G. Ikeorgu, & E. C. Nwachukwu	High yielding, pest and disease tolerant, very good for fufu, frying, and boiling.	2009	2009
NRCRI, Umudike	R. Asiedu, J. G. Ikeorgu, & E.C. Nwachukwu	High yielding, pest and disease tolerant, very good for fufu, frying, and boiling.	2009	2009

CROP NAME	S/N	VARIETY NAME	ORIGINAL NAME	NATIONAL CODE	ORIGIN/SOURCE
Yam	432	TDa 00/00194	TDa 00/00194	NGDA-09-15	IITA, Ibadan
Yam	433	TDa 00/00104	TDa 00/00104	NGDA-09-16	IITA, Ibadan
Yam	434	UMUDa-4	TDa 00/00364	NGDA-10-17	IITA, Ibadan
Yam	435	UMUDr-17	TDr 95/19177	NGDR-10-18	IITA, Ibadan
Yam	436	UMUDr-18	TDr 89/02475	NGDR-10-19	NCRI, Umudike

Source: Data from NACGRAB, *Catalogue of crop varieties released and registered in Nigeria*, vol. 4 (Ibadan: National Centre for Genetic Resources and Biotechnology, 2012).

DEVELOPING INSTITUTE	BREEDER/ COLLABORATING SCIENTISTS	OUTSTANDING CHARACTERISTICS	YEAR OF RELEASE	YEAR OF REGISTRY
NRCRI, Umudike	R. Asiedu, C. N. Egesi, & J. G. Ikeorgu	High yielding, pest and disease tolerant, good for pounded yam, frying, and boiling.	2009	2009
NRCRI, Umudike	R. Asiedu, C. N. Egesi, & J. G. Ikeorgu	High yielding, pest and disease tolerant, good for pounded yam, frying, and boiling.	2009	2009
NRCRI, Umudike	R. Asiedu, C. N. Egesi, & J. G. Ikeorgu	High yielding, good for *amala*, pounded yam, frying, and boiling.	2010	2010
NRCRI, Umudike	R. Asiedu, E. C. Nwachukwu, & J. G. Ikeorgu	High yielding under dry-season yam cropping system.	2010	2010
NCRI, Umudike	R. Asiedu, E. C. Nwachukwu, & J. G. Ikeorgu	High yielding, pest and disease tolerant, very good for fufu, frying, and boiling.	2010	2010

Notes

Introduction

1. FAOSTAT stands for Food and Agriculture Organization of the United Nations Statistics.

Chapter 1. Yam Primer

1. NRI 2005; Osei 1998; Oluwatusin 2011; Orkwor et al. 1996; Orkwor and Asadu 1998.
2. Henk Mutsaers presents an excellent discussion of yam production methods in another environment in Nigeria (Mutsaers 2007).

Chapter 2. Yam Production Contexts in Nigeria and Ghana

1. Members of the community were assembled and requested to group themselves into three by size of their yam production operations: large, medium and small. In each group one farm household was selected randomly.
2. Some African villages have market centers that meet periodically, such as once a week. The closer a village is to an urban center the more likely it is to have a periodic market center.
3. Dr. C. C. Okonkwo, personal communication, February 2014.
4. The farmer attitude toward trading in seed yam, which is not applicable to other staple crops, is a result of the high values of yam and the high cost of seed yam.

5. Issahaq Suleman, personal communication, February 2014.

6. Patrick Ngody, personal communication, November 2010.

7. "Suitable" in terms of high fertility soil, the low incidence of yam pests and diseases, and the availability of yam stake trees.

8. The numbers were recalculated from 70 percent after discounting 30 percent for seed.

9. Unagreed commission is the amount above the agreed commission the middleman may withhold from the proceeds because the farmer does not know how much the yam is sold for.

Chapter 3. High Labor Cost in Yam Production

1. One person day is eight working hours.

2. The practice is destructive in terms of rapid deforestation; yam is not grown in the same field two consecutive seasons because of accumulation of its soilborne pests and diseases.

3. Bagamsah and Osei 2001; Ikeorgu, Nwokocha, and Ikwelle 2000; Orkwor et al. 2000; Osiru and Hahn 1994; Bai and Ekanayake 1998.

4. Otoo and Lamptey 2002; Ikeorgu and Igwilo 2000; Ndegwe et al. 1990; Obiazi 1995; Orkwor and Asadu 1998; Osiru and Hahn 1994; Out and Agboola 1991; Nwachukwu and Obi 1999; Ndegwe 1992; Anuebunwa 1994; Igwilo 1992, 1998b, 1994; and IITA 1974.

Chapter 4. Yam Seed Technologies

1. Norbert Maroya, personal communication, 2012.

2. In the representative countries, there are no private consumer protection agencies and the legal systems are weak. The public sector regulatory agencies, not the consumer, hold the producer responsible for the label on the product.

3. Aighewi, Akoroda, and Asiedu 1995; Okonkwo and Okezie 1993; Aighewi, Asiedu, and Akoroda 2001.

4. Periderm is the outer layer of bark; parenchyma is the primary tissue of plants composed of thin-walled cells that remain capable of cell division even when mature.

5. Emokaro and Law-Ogbomo 2008; Sey 1994; Abudulai and Quansah 2002; Dabels and Ikeorgu 1999; Igwilo 1998a, 1999; Ikeorgu, Ezulike, and Nwauzor 1998; Ikeorgu and Aniedu 1999.

6. Ikeorgu, Anioke, and Nwauzor 1998; IITA 1983; Ikeorgu, Nwokocha, and Ikwelle 2000.

7. Abudulai and Quansah 2002; Enyinnia and Emehute 1998; Emehute, Orkwor, and Anioke 1996; Emehute 1999.

8. Emokaro and Law-Ogbomo 2008; Asumugha and Obiechina 2001; Ezeh 1996, 1991.

9. NRI 2005; Aidoo et al. 2011; Ntow 2008; Asumugha and Chinaka 1998; Asiabaka 1992; Chikwendu, Chinaka, and Omotayo 1995; Erhabor, Omoregie, and Idachaba 1994; Akoroda 1992; Maduekwe, Ayichi, and Okoli 2000.

Chapter 5. The Challenge of Nematodes in the Yam Food Sector

1. Hyperplasia is abnormal increase in number of cells; hypertrophy is abnormal enlargement of a body part or organ.
2. Gibson et al. 2009.
3. Chiedozie Egesi, email message, December 2013.
4. Regina Kapinga, personal communication, June 2012.
5. Djana Mignouna, personal communication, November 2013.

Chapter 6. Yam Crop Improvement Research

1. Information is not readily available on the history of yam research in francophone West Africa.
2. Robert Asiedu, personal communication, September 2013.

Chapter 7. Prospects and Impediments to Diffusion of Hybrid Yams in Nigeria

1. Drs. G. C. Orkwor, R. Asiedu, and S. Hahn are among the breeders who developed the hybrid yams.
2. Robert Asiedu, personal communication, May 2013.
3. The farmer group could not name the pest or disease.
4. The West African local names of the water yam variety are derived from the variety's rapid seed multiplication attribute and from the name of one of the diffusion agencies. For example, *asana*, a Hausa language word for matches, is given because seed yams of the variety as small as a match box produced high yield. The name *sudan*, which is used for the variety in Nigeria, is derived from the name of the Sudan United Mission that promoted the diffusion in Nigeria.
5. Information on promotion was provided by Dr. Chris C. Okonkwo.
6. See Abudulai and Quansah 2002; Adekayode 2004; Aighewi, Akoroda, and Asiedu 1995; Ano, Asiegbu, and Udealor 1999; Asare-Bediako et al. 2007; Asiabaka 1992; Asumugha and Chinaka 1998; Asumugha and Obiechina 2001; Beckford 2009; Chikwendu, Chinaka, and Omotayo 1995; Dabels and Ikeorgu 1999; Emehute, Orkwor, and Anioke 1996; Emokaro and Law-Ogbomo 2008; Enyinnia and Emehute 1998; Ezeh 1991, 1996, 1998; Igwilo 1998a, 1999; Ikeorgu and Aniedu 1999; Ikeorgu, Anioke, and Nwauzor 1997, 1998; Ikeorgu, Nwokocha, and Ikwelle 2000; Kalu and Erhabor 1992; Meyen, Basssey, and Ibedu

1995; Nwauzor 1996, 1998a; Odigboh and Akubuo 1991; Oguntade, Thompson, and Ige 2010; Okonkwo 1995; Orkwor et al. 1996; Sey 1994; Yankey 2002.

7. Ezeh 1991, 1996, 1998; Asiabaka 1992; Asumugha and Chinaka 1998; Asumugha and Obiechina 2001.

Chapter 8. Periphery Situation of Yam in National Food Policy Programs

1. Information is not readily available for amounts of seed and credit made available.
2. Data was unavailable for the period before 2002.
3. Information provided by Dr. Chris Okonkwo, February 2014.
4. School leavers refer to people who leave school for any reason, such as graduation or dropout.
5. "Welcome to the Ghana Rice Wars: Subsidized Imports versus Local Brands," *Ghana Business and Finance* (February 2011), http://www.ghanabizmedia.com/ghanabizmedia/february-2011-trade/225-welcome-to-the-ghana-rice-wars-subsidised-imports-versus-local-brands.html.
6. Price data are not readily available for cowpea.
7. Issahaq Suleman, personal communication, January 2014.
8. Ibid.

Chapter 9. Yam Stereotype as a Man's Crop

1. The Lower Niger is the Niger basin from just above the Niger Delta in the south to the confluence of the Rivers Niger and Benue at Lokoja in the middle of Nigeria.

Chapter 10. Yam Postharvest Matters

1. Literature did not provide evidence that this information has been updated in another study.
2. The next three paragraphs contain some information presented elsewhere in this book; this is necessary to emphasize the problems posed by serious deficiencies of the yam marketing system.
3. FAOSTAT information on yam Food Balance Sheets is less than credible.
4. As was done at the Global Yam Conference, October 2013, Accra, Ghana, and a YIIFSWA Progress Review and Work Planning Meeting, February 2014, IITA, Ibadan, Nigeria.

Chapter 11. Trade in West African Yam

1. A product has low added value if processed minimally.
2. R^2=85 percent in polynomial of order 2, the best fit model. R^2 of 85 percent means that

the trend is explained largely by time, and polynomial of order 2 means that the trend is variable.

3. R^2=66 percent with polynomial of order 2, the best fit model. R^2 of 66 percent means that the trend is explained largely by time, and polynomial of order 2 means that the trend is variable.

Chapter 12. Fuels of Yam Consumption in West Africa

1. Compared with Ghana, per capita yam consumption is low in Nigeria because the national average is misleading.
2. Apart from sorghum and millet, which are staples in a part of the country with a different cultural background.
3. A common African food is called *foofoo* in Anglophone West Africa, *foutou* in Francophone West Africa, and variously called *ugali*.
4. Maize is in dry form; the price differential will be wider if yam and cassava are converted to dry forms.
5. The association is stronger than the figures presented in the graphs portray because classification of respondents into income groups was subjective. The classification was not unique but varied with different enumerators.

Chapter 13. Yam in the Igbo Cultural Rite of Marriage

1. Here a medal of honor is the prize to a winner in a competition.

Appendix 1. Participants in Activities that Contributed Information Reported in this Book

1. List was provided by Folake Olokun.

References

Abraham, R. C. 1940. *The Tiv people.* London: Crown Agents.

Abudulai, Mumuni, and Charles Quansah. 2002. "Alternative media to sawdust for minisett propagation of seed yam (*Dioscorea* spp.)." *Tropical Science* 42(2): 47–51.

Achebe, Chinua. 1988. *Hopes and impediments: Selected essays, 1965–1987.* Oxford: Heinemann.

———. 2012. *There was a country: A personal history of Biafra.* New York: Penguin.

Acholo, M., S. Morse, N. Macnamara, L. Flegg, and R. P. Oliver. 1997. "Aetiology of yam (*Dioscorea rotundata*) tuber rots held in traditional stores in Nigeria: Importance of *Fusarium* spp. and yam beetle." *Microbiological Research* 152(3): 293–98.

Adekayode, F. O. 2004. "The economics of seed yam production by the yam minisett technique in a humid tropical region." *Journal of Food Technology* 2: 284–87.

Aidoo, Robert, Fred Nimoh, John-Eudes Andivi Bakang, Kwasi Ohene-Yankyera, Simon Cudjoe Fialor, and Robert Clement Abaidoo. 2011. "Economics of small-scale seed yam production in Ghana: Implications for commercialization." *Journal of Sustainable Development in Africa* 13(7).

Aighewi, B. A. 1998. "Seed yam (*Dioscorea rotundata* Poir.) production and quality in selected yam zones of Nigeria." PhD diss., University of Ibadan.

Aighewi, B. A., M. O. Akoroda, and R. Asiedu. 1995. "Preliminary studies of seed yam production from minisetts with different thickness of cortex parenchyma in white yam

(*Dioscorea rotundata* Poir.)." In *Root crops and poverty alleviation*, ed. M. O. Akoroda and I. J. Ekanayake, 445–47. Proceedings of the Sixth Triennial Symposium of the International Society for Tropical Root Crops–Africa Branch, Lilongwe, Malawi, October 22–28.

Aighewi, B. A., R. Asiedu, and M. O. Akoroda. 1998. "Producing seed yams from sprouts." In *Root crops in the 21st century*, ed. M. O. Akoroda and J. M. Ngeve, 297–99. Proceedings of the Seventh Triennial Symposium of the International Society for Tropical Root Crops–Africa Branch, Centre International des Conférences, Cotonou, Benin, October 11–17.

Akem, C. N., and R. Asiedu. 1992. "Distribution and severity of yam anthracnose." In *Root crops for food security in Africa*, ed. M. Akoroda, 297–311. Proceedings of the Fifth Triennial Symposium of the International Society for Tropical Root Crops–Africa Branch, Kampala, Uganda, November 22–28.

Akoroda, M. O. 1992. "A century of yam research in Nigeria: 1893–1992." In *Root crops for food security in Africa*, ed. M. Akoroda, 39–43. Proceedings of the Fifth Triennial Symposium of the International Society for Tropical Root Crops–Africa Branch, Kampala, Uganda, November 22–28.

Amikuzuno, Joseph. 2001. "Evaluating the efficiency of the yam marketing system in Ghana." MPhil thesis, Department of Agricultural Economy and Farm Management, University of Ghana, Legon.

Ano, A. O., J. A. Asiegbu, and A. Udealor. 1999. "Effect of green mulch of multipurpose tree species applied to yam minisett on soil physico-chemical properties and seed yield." In *Annual report*, 44–47. Umudike: NRCRI.

Anuebunwa, F. O. 1994. "On-farm evaluation of yam staking material alternatives in a yam-cassava based cropping system in the forest-savanna mosaic belt of Nigeria." *Biological Agriculture and Horticulture* 10: 179–88.

Asante, S. K., Esther Wahaga, and Issah Ramat. 2002. "Farmers' knowledge and perceptions of yam pests and their indigenous control practices in northern Ghana." Savanna Agricultural Research Institute, Tamale, Ghana.

Asare-Bediako, Elvies. 2002. "Studies of microorganisms affecting sprouting of mini yam setts of white yam (*D. rotundata* Ta poir) cv Pona in Ghana." Crp Research Institute, Kumasi, Ghana. 125 pp. MPhil thesis, University of Cape Coast.

Asare-Bediako, Elvies, F. A. Showemimo, Y. Opoku-Asiama, and D. H. A. K. Amewowor. 2007. "Improving sprouting ability of white yam minisetts (*Dioscorea alata* Poir.) varieties pona and dente using different disinfectants and protectants in sterilized saw dust." *Journal of Applied Science* 7(20): 3131–34.

Asiabaka, C. C. 1992. "The attitude of farmers towards the yam minisett technology in Imo State of Nigeria." In *Root crops for food security in Africa*, ed. M. Akoroda, 372–74.

Proceedings of the Fifth Triennial Symposium of the International Society for Tropical Root Crops–Africa Branch, Kampala, Uganda, November 22–28.

Asumugha, G. N., and C. C. Chinaka. 1998. "Socio-economic and cultural variables influencing adoption of yam minisett technique in three Eastern States of Nigeria." In *Sustainable agricultural development in a changing environment*, ed. M. C. Igbokwe et al., 56–59. Proceedings of the Thirty-Second annual conference of the Agricultural Society of Nigeria, Federal College of Agriculture, Ishiagu, Ebonyi State, September 13–16.

Asumugha, G. N., and C. O. B. Obiechina. 2001. "Comparative economics of minisett and traditional seed yam production technology at the farm level in the eastern forest zone of Nigeria." *African Journal of Root and Tuber Crops* 4(2): 9–12.

Bagamsah, T. T., and C. Osei. 2001. Further improvement in the gene pool. Savanna Agricultural Research Institute, Tamale, Ghana.

Bai, K. V., and I. J. Ekanayake. 1998. "Taxonomy, morphology and floral biology." In *Food Yams: Advances in Research*, ed. G. C. Orkwor, R. Asiedu, and I. J. Ekanayake, 13–37. Ibadan: IITA/NRCRI.

Baumann, Hermann. 1928. "The division of work according to sex in African hoe culture." *Africa: Journal of the International African Institute* 1(3): 289–319.

Beckford, Clinton L. 2009. "Sustainable agriculture and innovation adoption in a tropical small-scale food production system: The case of yam minisetts in Jamaica." *Sustainability* 1(1): 81–96.

Boserup, Ester. 1970. *Women's role in economic development*. New York: St. Martin's Press.

Boutillier, J. L. 1960. *Bonguoannou, Côte d'Ivoire: Étude socio-économique d'une subdivision.* Homme d'outre-Mer 2. Paris: Berger-Levrault.

Chikwendu, D. O., C. C. Chinaka, and A. M. Omotayo. 1995. "Adoption of minisett technique of seed yams production by farmers in the eastern forest zone of Nigeria." *Discovery and Innovation* 7: 367–75.

Chiwona-Karltun, Linley. 2001. "A reason to be bitter: Cassava classification from the farmers' perspective." PhD Thesis, Karolinska Institutet, Stockholm, Sweden.

Cook, Joelle, and Sara Curran. 2009. "Gender and cropping: Yam in Sub-Saharan Africa." Prepared for the Science and Technology Team of the Bill and Melinda Gates Foundation.

Coursey, D. G. 1967. *Yams: An account of the nature, origins, cultivation and utilisation of the useful members of the Dioscoreaceae*. London: Longmans.

Coyne D., A. Claudius-Cole, and H. Kikuno. 2010. "Sowing the seeds of better yam." CGIAR SP-IPM Technical Innovation Brief, No. 7, November.

Coyne, D. L., J. M. Nicol, and B. Claudius-Cole. 2007. *Practical plant nematology: A field and laboratory guide*. SP-IPM Secretariat, International Institute of Tropical Agriculture,

Cotonou, Benin.

Dabels, V. Y., and J. E. G. Ikeorgu. 1999. "Germination trial of yam minisetts using different growth media." In *Annual report*, 27. Umudike: NRCRI.

Degras, L. 1993. *The yam: A tropical root crop*. London: Macmillan.

Dennett, R. E. 1910. *Nigerian studies*. London: Macmillan.

Dorosh, Paul. 1988. "The economics of root and tuber crops in Africa." RCMP Research Monograph No. 1. Resource and Crop Management Program. Ibadan: IITA.

Doumbia, Sékou, Muamba Tshiunza, Eric Tollens, and Johan Stessens. 2004. "Rapid spread of the Florido yam variety (*Dioscorea alata*) in Ivory Coast: Introduced for the wrong reasons and still a success." *Outlook on Agriculture* 33(1): 49–54.

Ellis, A. B. 1890. *The Ewe-speaking peoples of the slave coast of West Africa*. London: Chapman and Hall.

———. 1894. *The Yoruba-speaking people of the slave coast of West Africa*. London: Chapman and Hall.

Emehute, J. K. U. 1999. "Studies on the shelf life of local plant-based yam minisett dust." In *Annual report*, 47–48. Umudike: NRCRI.

Emehute, J. K. U., G. C. Orkwor, and S. C. Anioke. 1996. "Sourcing of local raw materials as substitute(s) for yam minisett dust." In *National Agricultural Research Project (NARP), priority research and REFILS project report*, 13–18. Umudike: NRCRI.

Emokaro, C. O., and K. E. Law-Ogbomo. 2008. "The Influence of minisett size on the profitability of yam production in Edo State, Nigeria." *Research Journal of Agriculture and Biological Sciences* 4(6): 672–75.

Ene, L. S. O., and O. O. Okoli. 1985. "Yam improvement: Genetic considerations and problems." In *Advances in yam research: The biochemistry and technology of the yam tuber*, ed. Godson Osuji. Enugu: Biochemical Society of Nigeria in collaboration with Anambra State University of Technology.

Enyinnia, T., and J. K. U. Emehute. 1998. "On-station validation of new dust formulation for dressing yam minisetts for enhanced establishment." In *Annual report 1997 and programme of work for 1998*, 47. Umudike: NRCRI.

Erhabor, P. O., E. M. Omoregie, and F. S. Idachaba. 1994. "The state of agricultural research and its commercialization in Nigeria." *Tropenlandwirt* 95: 173–83.

Ezedinma, C., I. A. Ojiako, R. U. Okechukwu, J. Lemchi, A. M. Umar, L. Sanni, M. Akoroda, F. Ogbe, E. Okoro, G. Tarawali, and A. Dixon. 2007. *The cassava food commodity market and trade network in Nigeria*. Ibadan: IITA.

Ezeh, N. O. A. 1991. "Economics of seed-yam production from minisetts in Umudike in Southeastern Nigeria. Implications for commercial growers." In *Proceedings of the ninth symposium of the International Society for Tropical Root Crops*, ed. F. Ofori and S. K. Hahn, 378–81, Accra, Ghana, October 20–26.

———. 1996. "Benefit-cost analysis of improvements in the productivity of yam minisett technique." In *National Agricultural Research Project (NARP), priority research and REFILS projects report*, 19–21. Umudike: NRCRI.

———. 1998. "Economics of production and postharvest technology." In *Food Yams: Advances in Research*, ed. G. C. Orkwor, R. Asiedu, and I. J. Ekanayake, 187–214. Ibadan: IITA/NRCRI.

FAO (Food and Agriculture Organization of the United Nations). 2014. "Agricultural growth in West Africa (AGWA): Market and policy drivers." Rome: FAO.

Forde, D. 1964. *Yakö studies*. Oxford: Oxford University Press.

Galletti, R., K. D. S. Baldwin, and I. O. Dina. 1956. *Nigerian cocoa farmers: An economic survey of Yoruba cocoa farming families*. Oxford: Oxford University Press.

Ghana. 1964. *Seven year development plan 1963–64 to 1969–70*. Accra: Planning Commission.

———. 1971. *One year development plan, July 1970 to June 1971*. Accra: Planning Commission.

———. 2009. Food and Agriculture Sector Development Policy (FASDEP II). Ministry of Food and Agriculture, Accra.

———. 2010. Medium Term Agriculture Sector Investment Plan (METASIP). Ministry of Food and Agriculture, Accra.

Gibson, R. W., R. O. M. Mwanga, S. Namanda, S. C. Jeremiah, and I. Barker. 2009. *Review of sweet potato seed systems in East and Southern Africa*. International Potato Center (CIP), Lima, Peru. Integrated Crop Management, Working Paper 2009-1, 48 pp.

Green, K. R. 1995. Distribution and severity of foliar diseases of yam (*Dioscorea* spp.) in Nigeria. In *Root crops and poverty alleviation*, ed. M. O. Akoroda and I. J. Ekanayake, 439–44. Proceedings of the Sixth Triennial Symposium of the International Society for Tropical Root Crops–Africa Branch, Lilongwe, Malawi, October 22–28.

Hahn, S. K., D. S. O. Osiru, M. O. Akoroda, and J. A. Otoo. 1987. "Yam production and its future prospects." *Outlook on Agriculture* 16(3): 105–10.

Igwilo, Ndubisi. 1992. "Effect of height of stakes and intercropping with Okra (Abelmoschus esculentus) on the yield of yam (*Dioscorea* spp.) grown from seed yams." *Beitrage zur Tropischen Landwirtschaft und Veterinarmedizin* 30(4): 373–79.

———. 1994. "Effect of staking and population of yam and removal of maize leaf and ear in a yam/maize: Dry matter production and the mechanism of maize interference with yam yield." *Nigerian Agricultural Journal* 27(1): 73–82.

———. 1998a. "Field performance of yam (*Dioscorea* spp.) pieces in relation to surface area of periderm and sett thickness." *Nigerian Agricultural Journal* 29: 78–79.

———. 1998b. "Yield of yam tubers grown from minisetts in relation to height of stakes and intercropping with okra (*Abelmoschus esculentus Moench*) in the rainforest zone of Nigeria." *Nigerian Agricultural Journal* 29: 95–105.

———. 1999. "Effect of progressive removal of ground tissue with increasing plant population on the growth and yield of yam (*Dioscorea* spp.) grown from minisetts." *Nigerian Agricultural Journal* 30: 19–31.

IITA (International Institute of Tropical Agriculture). 1974. *Annual report.* Ibadan: IITA.

———. 1983. *Annual report.* Ibadan: IITA.

———. IITA. 1989. Strategic Planning 1986/87., Root and Tuber Improvement Program.

———. IITA. 1995. Yam Research at IITA 1971–1993. Crop Improvement Division, Root and Tuber Improvement Program. IITA, Ibadan.

———. 1996. *Annual report.* Ibadan: IITA.

———. 2004. *Annual report.* Ibadan: IITA.

———. 2007. *Annual report.* Ibadan: IITA.

Ikeorgu, J. E. G., and O. C. Aniedu. 1999. "Effect of variety and size of minisetts on yield and size of minituber." *Annual report,* 23–26. Umudike: NRCRI.

Ikeorgu, J. E. G., S. C. Anioke, and E. C. Nwauzor. 1997. "Investigations on component crops for intercropping compatibility with yam minisetts." In *Annual report 1997 and programme of work for 1998,* 37–39. Umudike: NRCRI.

———. 1998. "Effect of tissue on introduction of minisetts in minisett/maize/melon intercrop on yields of component crops." In *Annual report 1997 and programme of work for 1998,* 40–42. Umudike: NRCRI.

Ikeorgu, J. E. G., T. O. Ezulike, and E. C. Nwauzor. 1998. "Effect of sett size on the seed yam size." In *Annual report 1997 and programme of work for 1998,* 39–40. Umudike: NRCRI.

Ikeorgu, J. E. G., and Ndubisi Igwilo. 2000. "Development of off-season intercropped yam/vegetable production techniques in Nigeria." In *Annual report,* 16–17. Umudike: NRCRI.

Ikeorgu, J. E. G., H. N. Nwokocha, and M. C. Ikwelle. 2000. "Seed yam production through the yam minisett technique: recent modifications to enhance farmers' adoption." In *Potential of root crops for food and industrial resources,* ed. Makoto Nakatani and Katsumi Komaki, 372–75. Proceedings of the Twelfth Symposium of the International Society for Tropical Root Crops, Tsukuba, Japan, September 10–16, 2000.

IMF (International Monetary Fund). 2014. *World Economic Outlook (WEO) update: Is the tide rising?* January. Washington, DC: IMF.

Johnson, Michael E., and William A. Masters. 2004. "Complementarity and sequencing of innovations: New varieties and mechanized processing for cassava in West Africa." *Economics of Innovation and New Technology* 13(1): 19–31.

Johnston, Bruce F. 1958. *The staple food economies of western tropical Africa.* Stanford: Stanford

University Press.

Jones, W. O. 1972. *Marketing Staple Foods in Tropical Africa*. Ithaca, NY: Cornell University Press.

Kaberry, Phyllis Mary. 1952. *Women of the grassfields: A study of the economic position of women in Bamenda, British Cameroons*. London: H. M. Stationery Office.

Kalu, B. R., and P. O. Erhabor. 1992. "Production and economic evaluation of white guinea yam minisett under ridge and bed production system in a tropical guinea savanna location, Nigeria." *Tropical Agriculture, Trinidad* 61:78–81.

Krapa, N. Herbert. 2006. "Host status of yam component crops to *Scutellonema bradys*." BSc thesis, Department of Crop and Soil Science, Kwame Nkuruma University of Science and Technology.

Maduekwe, Michael C., Damian Ayichi, and Ernest C. Okoli. 2000. *Issues in yam research*. ATPS Working Paper No. 21. Nairobi: African Technology Policy Network.

Mahony, Frank, and Pensile Lawrence. 1959. "Yam cultivation in Ponape." Anthropological Working Papers, No. 4. Guam: Office of the Staff Anthropologist, Trust Territory of the Pacific Islands.

Maroya, Norbert, Morufat Balogun, Robert Asiedu, Beatrice Aighewi, P. Lava Kumar, and Joao Augusto. 2014. "Yam propagation using 'aeroponics' technology." *Annual Research & Review in Biology* 4(24): 3894–3903.

Meek, C. K. 1937. *Law and authority in a Nigerian Tribe: A study in indirect rule*. Oxford: Oxford University Press.

Me-Nsope, Nathalie M., and John M. Staatz. 2013. *Trends in per capita food availability in West Africa, 1980–2009*. International Development Working Papers 130. East Lansing: Michigan State University.

Meyen, I. W., A. D. E. U. Bassey, and M. A. Ibedu. 1995. "On-farm determination of the optimum rate of NPK fertilizer application to yam minisett intercropped with maize followed by cowpea." In *Proceedings of the 9ᵗʰ annual Farming Systems Research and Extension Workshop in South Eastern Nigeria*, 57–58. National Root Crops Research Institute, Umudike, Nigeria.

Miege, J. 1954. "Lescultures vivrieres en AfriqueOccidentale." *Cahiers d'Outre-Mer* 7(25): 25–50.

Miege, J., and S. N. Lyonga, eds. 1982. *Yams*. Oxford: Clarendon Press.

Mignouna, Djana, Adebayo Akinola, Issacq Suleman, and Felix Nweke. 2014. "Yam: A Cash Crop in West Africa." YIIFSWA Working Paper No. 3. IITA, Ibadan.

Mignouna, J., R. Mank, T. Ellis, N. van den Bosch, R, Asiedu, S. Ng, and J. Peleman. 2002. "A genetic linkage map of Guinea yam (*Dioscorea rotundata* Poir.) based on AFLP markers." *Theoretical and Applied Genetics* 105(5): 716–25.

Morse, S., M. Acholo, N. McNamara, and R. Oliver. 2000. "Control of storage insects as a means of limiting yam tuber fungal rots." *Journal of Stored Products Research* 36:37–45.

Mozie, O. 1996. "Effect of mechanical injury and regulated air flow on storage losses of white yam tubers." *Tropical Science* 36(2): 65–67.

Mutsaers, H. J. W. 2007. *Peasants, farmers and scientists: A chronicle of tropical agricultural science in the twentieth century.* Dordrecht: Springer.

NACGRAB (National Centre for Genetic Resources and Biotechnology). 2012. *Catalogue of crop varieties released and registered in Nigeria.* Vol. 4. Ibadan: National Centre for Genetic Resources and Biotechnology.

Ndegwe, N. A. 1992. "Economic returns from yam/maize intercrops with various stake densities in a high-rainfall area." *Tropical Agriculture* 69(2): 171–75.

Ndegwe, N. A., F. N. Ikpe, S. D. Gbosi, and E. T. Jaja. 1990. "Effect of staking method on yield and its components in sole-cropped white Guinea yam (*Dioscorea rotundata* Poir.) in a high-rainfall area of Nigeria." *Tropical Agriculture* 67(1): 29–32.

Neuenschwander, Peter. 2003. "Biological control of cassava and mango mealybugs in Africa." In *Biological control in IPM systems in Africa*, ed. P. Neuenschwander, C. Borgemeister, and J. Longewald, 45–59. Oxford: CABI.

Nigeria. 2011. *Agricultural transformation agenda: We will grow Nigeria's agricultural sector.* Abuja: Federal Ministry of Agriculture and Rural Development.

NRI (Natural Resources Institute). 2005. *Evaluation and promotion of crop protection practices for "clean" seed yam production systems in Central Nigeria.* Crop Protection Programme. Greenwich: NRI, University of Greenwich.

Ntow, Frederick. 2008. "Yam minisett technology: Adoption and performance in the Nanumba North District." Savanna Agricultural Research Institute, Tamale, Ghana.

Nwachukwu, E. C., and I. U. Obi. 1999. "Clonal evaluation of MV$_2$ yam lines raised from gamma ray treated Obiaoturugo minitubers." In *Annual report*, 41–43. Umudike: NRCRI.

Nwankiti, A. O., and O. B. Arene. 1978. "Diseases of yam in Nigeria." *International Journal of Pest Management* 24(4): 486–94.

Nwankiti, Okechukwu Alphonso, and E. U. Okpala. 1980. "Anthracnose of water yam in Nigeria." In *Tropical root crops: Research strategies for the 1980s*, ed. E. R. Terry, K. A. Oduro, and F. Caveness, 166–70. Proceedings of the First Triennial Root Crops Symposium of the International Society for Tropical Root Crops–Africa Branch, Ibadan, Nigeria, September 8–12.

Nwauzor, E. C. 1996. "Use of groundnut (*Arachis hypogaea* L.) as an intercrop for nematode control in seed yam production by minisett techniques final report." In *National Agricultural Research Project (NARP), priority research and REFILS projects report*, 22–26.

Umudike: NRCRI.

———. 1998a. "Evaluation of dust preparation from Neem and 'Otiri' for seed dressing as alternative to yam minisett dust." In *Annual report 1997 and programme of work for 1998*, 42–43. Umudike: NRCRI.

———. 1998b. "Nematode problems and solutions of root and tuber crops in Nigeria." In *Root crops in the 21st century*, ed. M. O. Akoroda and J. M. Ngeve, 545–631. Proceedings of the Seventh Triennial Symposium of the International Society for Tropical Root Crops–Africa Branch, Centre International des Conférences, Cotonou, Benin, October 11–17.

Nweke, Felix I., Robert Aidoo, and Benjamin Okoye. 2013. "Yam consumption patterns in West Africa." Unpublished report submitted to Bill and Melinda Gates Foundation.

Nweke, Felix I., Malachy Akoroda, and John Lynam. 2011. "Seed systems of vegetatively propagated crops in sub-Saharan Africa: Report of a situation analysis." Unpublished report prepared for Bill and Melinda Gates Foundation.

Nweke, Felix I., Dunstan S. C. Spencer, and John K. Lynam. 2002. *The cassava transformation: Africa's best-kept secret*. East Lansing: Michigan State University Press.

Nweke, Felix I., B. O. Ugwu, C. L. A. Asadu, and P. Ay. 1991. "Production costs in the yam-based cropping systems of southeastern Nigeria." RCMP Research Monograph No. 6. Resource and Crop Management Program, IITA.

Obiazi, C. C. 1995. "Sustainable supply of stakes for yam production." *Journal of Sustainable Agriculture* 5(3): 133–38.

Odigboh, E. U., and C. O. Akubuo. 1991. "A two-row automatic minisett yam planter." *Journal of Agricultural Engineering Research* 50:189–96.

Odu, B. O., S. A. Shoyinka, J. d'A. Hughes, R. Asiedu, and A. O. Oladiran. 1998. "Yam viruses in Nigeria." In *Root crops in the 21st century*, ed. M. O. Akoroda and J. M. Ngeve, 631–33. Proceedings of the Seventh Triennial Symposium of the International Society for Tropical Root Crops–Africa Branch, Centre International des Conférences, Cotonou, Benin, October 11–17.

Oguntade, A. E., O. A. Thompson, and T. Ige. 2010. "Economics of seed yam production using minisett technique in Oyo State, Nigeria." *Field Actions Science Reports* 4. Http:// factsreports.revues.org/659.

Ohadike, D. C. 1981. "The influenza pandemic of 1918–19 and the spread of cassava cultivation on the Lower Niger: A study in historical linkages." *Journal of African History* 22:379–91.

Okoli, O. O. 1975. "Yam production from seeds: Prospects and problems." Paper presented at the eleventh annual conference of the Agricultural Society of Nigeria, Enugu, July 1–15.

———. 1980. "Parameters for selecting parents for yam hybridization." In *Tropical Root Crops: Research Strategies for the 1980s*, ed. E. R. Terry, K. A. Oduro, and F. Caveness, 163–65.

Proceedings of the First Triennial Root Crops Symposium of the International Society for Tropical Root Crops–Africa Branch, Ibadan, Nigeria, September 8–12.

Okoli, O. O., and M. O. Akoroda. 1995. "Providing seed tubers for the production of food yams." *African Journal of Root and Tuber Crops* 1(1): 1–6.

Okonkwo, J. C. 1995. "Effects of pre-sprouting and planting depth of yam (*Dioscorea* spp.) minisett on field establishment, tuber yield and net income in the southern guinea savannah of Nigeria." *Nigerian Agricultural Journal* 28(1): 1–8.

Okonkwo, S. N. C., and C. E. A. Okezie. 1993. "Potential non-tuber materials for yam propagation." In *Advances in yam research*, vol. 2, *Production and post-harvest technologies of the yam tuber*, 43–55. Nsukka: Department of Botany and Division of General Studies, University of Nigeria.

Okorji, E. C. 1983. "Consequences for agricultural productivity of crop stereotyping along sex lines: A case study of four villages in Abakaliki area of Anambra State." MSc thesis, Department of Agricultural Economics, University of Nigeria, Nsukka.

Olayide, S. O. 1981. *Scientific research and the Nigerian economy*. Ibadan: Ibadan University Press.

Oluwatusin, Femi Michael. 2011. "Measuring technical efficiency of yam Farmers in Nigeria: A stochastic parametric approach." *Agricultural Journal* 6(2): 40–46.

Onwueme, I. C. 1977. "Effect of varying the time of the first harvest, and of late planting, on double-harvest yield of yam (*Dioscorea rotundata*) in field plots." *Journal of Agricultural Science* 88: 737–41.

Orkwor, G. C., and C. L. A. Asadu. 1998. "Agronomy." In *Food yams: Advances in research*, ed. G. C. Orkwor, R. Asiedu, and I. J. Ekanayake, 105–41. Ibadan: IITA/NRCRI.

Orkwor, G. C., R. Asiedu, and I. J. Ekanayake, eds. 1998. *Food yams: Advances in research*. Ibadan: IITA/NRCRI.

Orkwor, G. C., R. Asiedu, S. K. Hahn, D. Surma, U. Udensi, and G. O. Chukwu. 2001. "Development and evaluation of hybrid yams (*Dioscorea rotundata* Poir.) in prerelease multilocational trials in Nigeria." In *Root crops: The small processor and development of local food industries for market economy*, ed. M. O. Akoroda. Proceedings of the Eighth Triennial Symposium of the International Society of Tropical Root Crops–Africa Branch, International Institute of Tropical Agriculture, Ibadan, Nigeria, November 12–16.

Orkwor, G. C., E. C. Nwachukwu, V. A. Dabels, and A. A. Adeniji. 2000. Multilocational evaluation of local land races and Hybrid yams. In *Annual report*, 4–5. Umudike: NRCRI.

Orkwor, G. C., O. O. Okoli, J. K. U. Emehute, and N. O. A. Ezeh. 1996. "Studies on the optimum plant population, depth of planting and best tuber portion of mother seed yam as planting sett in seed yam production using minisett technique." In *National Agricultural Research*

Project (NARP), priority research and REFILS projects report, 4–12. Umudike: NRCRI.

Osei, Cecil. 1998. "Effect of rate and time of application of organic and inorganic fertilizers on yield and quality of yam." Savanna Agricultural Research Institute.

Osei-Serpong, Kwadwo. Undated. "Effect of mother yam on minisett technique." Slide Presentation. Crop Research Institute Kumasi/Ministry of Food and Agriculture, Accra.

Osiru, D. S. O., and S. K. Hahn. 1994. "Effects of mulching materials on the growth development and yield of white yam." *African Crop Science Journal* (Uganda) 2(2): 153–60.

Otoo, E., R. Akromah, M. Koleniskova-Allen, and R. Asiedu. 2009. "Ethno-botany and morphological characterisation of the yam pona complex in Ghana." *African Crop Science Conference Proceedings*, 9:407–14.

Otoo, E., and J. N. L. Lamptey. 2002. "Evaluation of *Dioscorea* species for adaptation to non-staking in Ghana." Paper presented at the Root and Tuber Improvement Research Review Workshop, November, Soil Research Institute, Kumasi, Ghana.

Out, O. I., and A. A. Agboola. 1991. "The suitability of Gliricidia Sepium in-situ: Live stake on the yield and performance of white yam (*Dioscorea rotundata*)." In *Tropical root crops in a developing economy*, ed. F. Ofori and S. K. Hahn, 360–66. Proceedings of the Ninth Symposium of the International Society for Tropical Root Crops, Accra, Ghana, October 20–26.

Rattray, R. S. 1923. *Ashanti*. Oxford: Oxford University Press.

RTIMP (Root and Tuber Improvement and Marketing Program). 2009. Status report as of October 2009. Ministry of Food and Agriculture, Programme Coordination Office, Kumasi.

Sadik, Sidki, and O. U. Okereke. 1975. "A new approach to improvement of yam *Dioscorea rotundata*." *Nature* 254(5496): 134–35.

Scaglion, Richard. 2007. "Abelam: Giant yams and cycles of sex, warfare and ritual." In *Discovering Anthropology: Researchers at Work- Cultural Anthropology*, ed. C. R. Ember and M. Ember, 21–31. Upper Saddle River, NJ: Pearson Prentice Hall.

Segnou, A., C. A. Fotokun, M. O. Akoroda, and S. K. Hahn. 1992. "Studies on the reproductive biology of white yam (*Dioscorea rotundata* Poir.)." *Euphytica* 64(3): 197–203.

Sey, Samuel. 1994. "Evaluation of the optimum sett weight in the minisett technique of seed yam propagation." BSc thesis, Department of Crop and Soil Sciences, Kwame Nkuruma University of Science and Technology.

Soyinka, Wole. 2012. *Harmattan haze on an African spring*. Ibadan: Bookcraft.

Spencer, D. S. C. 1976. "African women in agricultural development: A case study in Sierra Leone." Overseas Liaison Committee Paper No. 9. Washington DC: American Council on Education.

Talbot, P. A. 1932. *Religion and art in Ashanti*. Oxford: Oxford University Press.

Tauxier, L. 1932. *Religion, moeurs et coutumes des Agnis de la Côte d'Ivoire*. Paris: Paul Geuthner.

Thomas, N. W. 1910. *Anthropological Report on the Edo-speaking peoples of Nigeria*. London: Harrison.

Thrupp, Lori Ann, Peter Veit, Clement Dorm-Adzobu, and Okyeame Ampadu-Agyei. 2006. "Environmental impact review of the non-traditional agricultural export sector in Ghana." Washington, DC: Center for International Development and Environment, World Resources Institute.

Tshiunza, Muamba. 1995. "Comparative study of production labor for selected tropical food crops (cassava, yam, maize and upland rice)." In *Root crops and poverty alleviation*, ed. M. O. Akoroda and I. J. Ekanayake, 121–25. Proceedings of the Sixth Triennial Symposium of the International Society for Tropical Root Crops–Africa Branch, Lilongwe, Malawi, October 22–28.

Ugwu, Boniface Omans. 1990. "Resource use and productivity in food crop production in major yam producing areas of southeast Nigeria." PhD diss., University of Nigeria, Nsukka.

Waitt, W. 1961. "Review of yam research in Nigeria, 1920–1961." Federal Department of Agricultural Research, Nigeria.

WFP (World Food Programme). 2009. "Republic of Ghana: Comprehensive food security and vulnerability analysis." May. Rome: World Food Programme.

Yankey, Egya Ndede. 2002. "The effects of growth regulators, yam part, on sprouting of *Dioscorea rotundata* Poir. variety 'Puna' minisetts in two pre-sprouting media." BSc thesis, Department of Crop and Soil Science, Kwame Nkuruma University of Science and Technology.

Index